Chemistry of Olive Oil

Calixto López

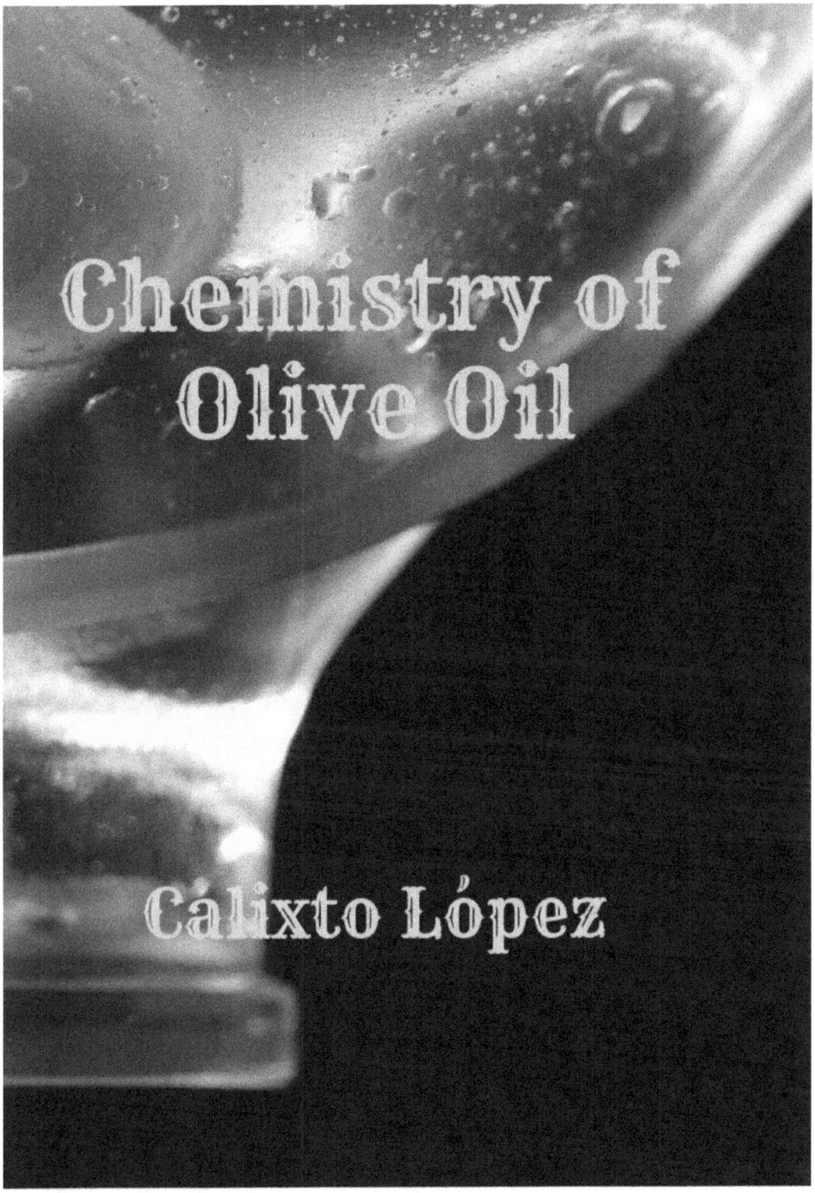

Chemistry of Olive Oil

Calixto López
(2019)

Chemistry of Olive Oil

Author's Foreword

Defenestrated in occasions, had to less in others, forgotten by some, always reborn like phoenix the one that for many is the king of the oils, the one that all pursue, the one that they try to equal and even adulterate, yes, **the olive oil**, the liquid of tenuous green-yellowish aspect that, without exception, is considered like the most valued of the vegetable oils.

That olive oil currently reigns in kitchens? Nobody dares to doubt, it the prices indicate it, also the most demanding tastes; it is considered the bow ship of the well-weighted Mediterranean cuisine, but why? What makes it so precious? Why do they all imitate it? Although none of them even equals it. What is it made of?

Each and every one of these questions will be answered in this book, which deals exclusively with this singular oil, and for this we will start from the most elementary and central part of everything: its *chemical composition*, hence its properties and, consequently, its use and applications.

The text will focus its attention on the basic components of olive oil, those that give it its properties and make it what it is today: *the prince of oils*, and that it can still fight in that bloody war waged in the world of vegetable oils. And he will have to enter the fight followed by a small army to confront the most powerful oils, against palm oil, sunflower oil, rapeseed oil, corn oil, cotton oil, and even with the apparently shy and fresh from the forest: avocado oil.

But it is not only against its kind that it fights the millennial

3

olive oil, with its weak flanks because of its nature dependent on centennial trees, perhaps tired of living after so many battles. It faces stoically the impetuous advances of technology, also against implacable plagues like the *Xylella fastidiosa,* and sometimes even against its creator: the man himself, not content with his work tries to demand more from what he cannot more, because his forces are not inexhaustible and are restricted by his own nature.

But olive oil has powerful weapons that until now have not been able to match their opponents, its *natural lipid profile rich in oleic acid,* without the need for changes or genetic transformations, and more than anything: its innocence, its virginity, like a vestal goddess before which all are forced to bow in respect, because this innocent creature is accompanied by more than 300 minority components many of them endowed with beneficial properties for the health of man, who is in the end who produces and consumes it.

Editor's Note:

In the translation of the book into the English language, in addition to human labour, translation programs were used. We are sorry if you find any grammatical errors in the text.

Subject Index

INTRODUCTION

Chemistry of Olive Oil

INTRODUCTION

There is no doubt that olive oil currently occupies the most prominent place among vegetable oils in terms of its nutritional properties and its health benefits. It is also the bow nave of the Mediterranean diet and its history parallel to that of the great ancient cultures of humanity, so that ancient ceramic vessels containing it are preserved, and the Greek and Roman classics frequently allude to its use and properties. Allegoric mosaics have also appeared for its use.

The olive branches adorned the head of the Roman emperors and the leaves are a symbol of peace and wellbeing, it is currently the best quoted oil on the market of the entire planet even though it does not occupy the first places in terms of volume of world production.

In spite of the above, difficult times are predicted and not very good auguries for this singular oil attending to a group of factors on which reference will be made through the exhibition of the contents of the book. It also has many enemies in sight among its peers, and its current production and trade is surrounded by uncertainties and contradictions, even among its growers and retailers.

Talking about olive oil is similar to talking about olives and olive trees, those hundred-year-old trees that swarm the countryside of southern Spain, Tuscany and other Italian regions, Greece, and the other countries of North Africa. Its production techniques have changed very little since ancient times, as well as the technification and mechanisation of crops, which favours the maintenance of small-scale artisans productions with techniques and methods handed down from generation to generation, generally accompanied by products of very high quality and commercial value, even related to protected trademarks of origin, although in the oil sector there is nothing virgin and protected, except the oil itself when it is not adulterated.

The basic problems for olive oil stem from its particular composition and the nature of the olive trees themselves. These trees are not capable of maintaining a stable production, so that in them the phenomenon of alternating good and bad harvests occurs indistinctly, which causes considerable differences in production volumes from one year to the next, with the upheavals and difficulties that this entails for production and trade.

The new olive tree plantations do not begin to produce immediately like those of their seed competitors, whose harvests are annual: sunflower, maize, rapeseed, peanuts, etc. So once the olive trees are planted, they have to wait a few years to produce olives, and more than that, to increase and stabilize fruit production up to an adequate exploitation volume.

On the other hand, at the moment the olive trees are being attacked by different diseases within which the one produced by the terrible *Xylella fastidiosa* stands out, that causes the drought, first of its leaves and branches, and later of the whole plant, as well as to spread the disease through the insects and to destroy entire plantations in very little time. In addition, as it attacks not only olive trees, but also citrus and other fruit and woody trees, the spectrum and possibilities of propagation are widened, as has happened in Italy in recent years with the destruction of considerable cultivation areas and the death of hundreds of thousands of trees, which has been a serious damage to the sector of Italian olive growers.

The bacteria have already entered peninsular Spain via the Balearic Islands and the fight is being waged to prevent their spread, because otherwise a sector that moves billions of euros and provides hundreds of thousands of jobs in the world's main olive oil producer would be significantly affected, so it is feared that this disease may reach huge areas of cultivation, although fortunately, this has not yet happened and measures are being taken to this effect.

The incidence of *Xylella fastidiosa* in crops, will be discussed later, when olive trees are studied.

Before falling in the competitors of the olive trees, it is necessary to emphasize that the star product of this agribusiness is the virgin olive oil, because the refined ones, when being submitted to different physicochemical processes, including heating, lose many of the beneficial properties of the virgin oil reason why they tend to differ less of their congeners extracted and refined of other plants.

This zeal in maintaining the seal of quality in olive oils in their virgin product, pays its share in the sense that the techniques and quality tests are based on the measurement of organoleptic indicators carried out by people, who although specialists in their genus, do not give the margins of reliability that the tests and instrumental analytical techniques of assessment, which includes a subjective element in the process, and the possibility, as indeed occurs, that frequent cases of adulteration occur given the high value of the product in the market and its wide demand. In such a way as to make it easier to "give a cat for a hare", as has been proven, and not infrequently. The adulteration of olive oil will also be dealt with in this monographic study and allusion will be made to some of the most notorious cases known in recent times, not only in Europe, but throughout the world, as it is an evil with which the olive oil industry has had to coexist throughout time.

The mechanization of olive groves, as they are trees that need to suffer as little damage as possible during the harvesting period, is not resolved as simply as in other sectors of seed oil production, where it is not necessary to preserve the plants at all, because the supplanting is from season to season, or from year to year, with seeds and new plants that perish at the end of the harvest, except for the African palm, but as this is for large bunches, the plants do not suffer appreciable damage and the cost of cultivation and maintenance is much lower.

But an even greater enemy weighs on the industry of the olive trees, and it is that of the competent oleaginous plants like the sunflower, the corn, the soybean and the colza among others, in what is related to the genetic modifications or transformations that are suffering their seeds to elevate the concentration of oleic acid in their composition and lipid profile, The paradigmatic element of olive oil, its high concentration of oleic acid, a beneficial element as a protector of cardiovascular diseases (**CVD**) is ceasing to be a monopoly of this oil and there is already frequent and natural talk of sunflower oil, corn, etc. high oleic acid, as well as oils for frying with high oleic acid content, which approaches, equals and may even surpass, the same oleic acid in olive oil.

Even the discriminated African palm with its new hybrid **OxG** with the American palm boasts an elevation of more than 10% of oleic acid in its composition and currently in South America the production figure of this oil exceeds hundreds of thousands of tons, with a constant expansion of the area of cultivation of this hybrid product. The oils competing with olive oil will be dealt with later, although palm oil is not yet distinguished in this field.

In order to argue a little more about the above, already in the Spanish market, the jealous guardian of olive oil, the sales levels of sunflower oil, to give an example, compete with those of the latter, bearing in mind that the price of 1 L of sunflower oil is four or five times lower than that of olive oil, and customers, of course, value their pockets which in times of crisis have not been very full, rather almost empty.

And this high price of the olive oil in elation with that of other vegetable oils, given the difficulties of the cultivation of the olive trees, the mechanization and especially its lower yield per hectare, cause that a prow industry tends to stagger in the countries of the Mediterranean basin. Fortunately, production is relatively low in relation to demand, which means that prices are still stable.

And to conclude in the description of this tidal wave of events affecting the world of olive groves, it is necessary to point out that apparent relatives of olive oil have appeared beyond the tropics that also threaten their kingdom, we refer to avocado oil, which boasts high levels of monounsaturated fatty acids such as oleic acid, at around 60%, and a minor relative, palmitoleic acid with a discrete 10%, but that adding a certain proportion of linoleic acid that contains make this oil rich in unsaturated fats make this oil rich in unsaturated fats, adding them to a certain proportion of linoleic acid that contains make this oil rich in unsaturated fats.

The avocado tree *(Persea americana)* is a frequent tropical plant in crops in South America and the Caribbean, and recently, for its properties, more the fruit than the oil, has expanded its exploitation to countries with subtropical climates such as Spain (South and Canary Islands) and has high productivity, There are varieties that adapt to different growing seasons, together with the characteristics of the plant, without thorns, and sizes that can surpass that of the shrubs, which makes it possible to establish a strong competition with the olive trees.

The avocado, of exquisite and soft mass, highly nutritious, is an ideal companion of the meals, sometimes it substitutes other viands, and in salads its use is very appreciated. To make matters worse, its virgin olive oil has a shade close to that of olive oil and is also accompanied by a high concentration of vitamins and antioxidants, which has led to cases of adulteration being detected from this oil as a component.

Under this prism the world of olive oil will be focused, a product with many years of walking and that still maintains its reign among the oils, but that as it has been seen, its dominion is strongly threatened.

The Prince of Oils

CHAPTER I

The Prince of Oils

Recently, when writing a chapter on olive oil in a book entitled *Chemistry of Vegetable Oils*, the author named it metaphorically "*The Prince of Oils*" and now this content is reflected again with slight modifications.

The reason why olive oil can currently be considered as *the prince of oils* is because this marvellous, lucid, transparent and beautifully coloured liquid combines the main nutritional and beneficial properties for the organism that shows any oil, at least until others appear, with attributes whose studies reveal that they are the most suitable at that moment, in the apparently slow becoming of history.

Oils are not people, but they are personalized through markets, through the regions that produce them, through their nutritional properties, and through man's need to subsist under certain conditions.

And nowadays olive oil with its soft yellowish green tonality, if it is refined, or darker if it is *virgin*, can be considered as the true prince who reigns over the other oils, and all given by its high content of oleic acid (**C18:9:1**), a monounsaturated fatty acid with the double bond in position 9 (**omega 9**) of its chain of 18 carbon atoms.

But it is not only the fact that olive oil contains high concentrations of oleic acid that makes it reign and reach the high prices that it presents in markets around the world, but also the substances that accompany it, especially in its raw state, in other times impurities such as antioxidant tocopherols, polyphenols, the vitamins it contains, its minerals and a whole group of components with proven benign properties for health.

But this was not always the case, there were times when olive oil was considered, due to its typical bitterness and roughness, an oil inferior to other more lucid and refined oils; but everything began to change as soon as clinical trials showed that it had the ideal concentration of fatty acids of different types: saturated, monounsaturated and polyunsaturated, without unbalancing the balance in one direction or another.

But the kingdom is in danger, and above all the monopoly on this type of oil presented by the countries of the Mediterranean Basin, its main producers, headed by Spain, followed by Italy and Greece as main exporters. It is also produced and elaborated from the earliest civilizations by the countries of North Africa that occupy the Mediterranean strip.

Soon, sooner or later, olive oil, *the prince of oils*, could change its residence and be located in North or South America, since countries such as the United States, Brazil, Argentina, Uruguay and Chile, among others they fight to gain a relevant position in its production and marketing. In Asia, the same could happen with China, or in Oceania with Australia, even in the rest of the African continent, where South Africa is emerging as a good producer of the precious liquid.

And if these countries have not yet achieved their expansionist aims in the olive oil production sector, it is not due to a lack of will or resources, much less extensions of arable land, nor due to commercial, technological and other barriers, but because their main opponent is olive trees: Those centenary trees that carry the **olives** with which *the green gold of the gods* is made, but as was said, it will be an unpredictable interval of time, more or less short or long, depending on the climatic conditions in which the olive trees develop in the countries of reference and the time in which they take to become adults.

Nowadays, olive oil has prices on the market several times higher than those of its seed competitors: sunflower, soy, rapeseed and maize, or those extracted from the fruit of the African palm, the most produced in the world. Because as a good prince, the production of olive oil is still expensive, limited and nothing comparable to that of the oils mentioned.

The aggressiveness of palm oils is notorious, in a few decades have been made with the largest volume of production in the world, and can not even compare with it's, closest competitors: soybeans and rapeseed.

Palm oil, thanks to its high productivity and its multiple uses in the food industry is the protected and favourite of the broad sector of jams, flour, fast food and many other spheres of food processing. Behind it are the giants of the food industry from all over the world, including Europe, who jealously watch over their protégé.

But what gives palm oil its great advantage in ensuring the finish and durability of jams, are the saturated fatty acids, but at the same time also attempt against it by the atherogenic damage associated with this type of fatty acids in human metabolism, by increasing serum cholesterol levels and accumulation in the arteries, thus hindering blood circulation and other associated ills.

Thus, there is little or no criticism of olive oil, perhaps the harshness or bitterness of its taste when it is extra virgin or virgin, but experts in the field consider that this is lack of culture in relation to this oil, that this is characteristic of its quality and sample of the beneficial substances it contains, such as polyphenols and for this, nothing, or very little has changed in the olive industry since time immemorial.

When the harvest of the olives arrives, they arrive from all humble parts humble day labourers to make their painful tasks, by means of little advanced techniques, because we are not talking about vast extensions of grasses, *astereaceaes, brasicaceaesu* or other cultures of small size and weak robustness, easily cut and threshed by means of harvesting machines; but of resistant trees in their full stature and adulthood, that have been silent witnesses of multiple historical events; of the promises of couples of lovers seated or hidden behind their trunks, of droughts, storms and storms, of the cold of the winter nights and of the intense heat of the Mediterranean summers.

The olive trees, those noble, resistant and perennial trees, do not understand much of mechanization and much less that they are being shaken with machinery, or that their fruit is not chosen sometimes in a selective way, reason why they do not welcome the technological advances, which makes that their cultivation is relatively expensive in comparison with that of other oleaginous plants, above all of the African palm their emulsion, for being oil of fruits.

And the miserable but self-sacrificing day labourers come at the time of harvesting olives from all over the world, perhaps even from distant regions, and dream of good harvests, and that next year's harvests will be better for employment and thus obtain what is necessary for the sustenance of their families; while they live in relatively precarious conditions, but without complaining or protesting, at least aloud, while their dreams rarely come true to the extent necessary, although perhaps that of the great producers, who for centuries have maintained names and brands recognized around the world.

The producers also have furtive enemies who, under the protection of the night, and perhaps without it, ignore the signs of ownership and take for themselves what is not their own, and steal some lots of olives, a deplorable fact, but apparently not very significant due to the high volume of harvests.

The olive oil continues being *the prince of the oils*, all struggle to equal it although they have to vary its lipid composition through not orthodox techniques, by means of the genetic engineering and the biotechnology, acquiring transgenic forms and sometimes even changing its name: oil of canola by rapeseed, sunflower and high oleic corn, among others.

Copying has its advantages and also its dangers, and not only because of the change of identity, but also because it is not known what can happen with the transgenic varieties of oil plants when it is done overnight what nature would have cost thousands of years to execute.

The oil war is complicated, so it is done for the olive oil, the current prince, there are many open fronts and day by day new opponents arise, so it does not reign all over the world: in Canada, Germany and some Central European countries, does the rapeseed (canola), in Ukraine and Russia the sunflower and corn, in the United States soybeans, corn and cotton, and in Brazil and Argentina occurs something similar to the American giant.

In Asia and the tropical countries of America, African palm and maize share dominance, although the former advances at an overwhelming pace, like a road roller in free fall, tamping and making everything disappear in its wake: plant and animal species, sustainable agriculture, and displacing human groups, some in a primitive tribal state, to uncertain destinations in the jungle of big cities.

The prince of oils faces his uncertain destiny, immutable, courageous, resigned and perhaps enjoying, ecstatic in his glory, as if to say that sentence grammatically vulgar and to some incoherent: and after they take away my dance, because more than being an oil, not a living being, it is understood that its greatness and decadence depends on man, the one who created it, the one who enjoys it, and the one who truly conducts the war of the oil market, in a chaotic, brutal and disproportionate way.

OLIVE

CHAPTER II

Olea europaea (Olive Tree)

The thousand-year-old olive tree

The olive tree, olive tree or *Olea europaea*, which is its scientific name and from whose fruits the olive oil is extracted, is an evergreen plant that reaches around 15 meters of height throughout its many years of life, because it can reach, and even surpass the century of existence, reason why its growth is relatively slow in comparison with other lasting trees like those of the tropical forests.

Olive trees need subtropical climates of moderate humidity, over 600-800 mm of precipitation per year.

The olive tree has a relatively thick trunk that allows it to give stability to its wide crown in the wind and other inclement weather. The shell, or bark, is widely fissured and is silvery grey.

The tree produces beautiful lancelet and pointed leaves that reach an average length of about 5 cm, although they may be shorter or longer. These were associated with peace and victory in antiquity and it is said that crowns made with them adorned the heads of Roman emperors. These are currently being studied by the scientific world attending to a group of polyphenols and other substances of biological interest that have been isolated from their extracts.

The flowers are hermaphrodites and the fruit, which is what most emphasizes its importance the well-known olive, reaches an average of 2.5 cm long. Initially it is green, but it tends to alter its colour until yellow and later dark purple in the stage of maturation.

The fruit of the olive tree, the olive, is extremely rich in oil in which monounsaturated fatty acids predominate, such as oleic acid, omega 9: (**C18:9:1**). It takes more than six months to mature.

In the Mediterranean Basin the plant flowers between the months of May and June, when spring arrives, to reach the time of ripening and harvesting of the fruit in the last months of the year.

The common olive tree has a wild younger brother, commonly known as acebuche, which tends more to be a shrub with similar fruits, but smaller. It is highly resistant to soil degradation and high temperatures, but not to intense cold. It shares fertility with the olive tree for what it must have contributed, in spontaneous crossings, its relatively high resistance to the attacks of the climate and the precariousness of the soil.

The olive tree is closely related to Mediterranean history and culture and is a typical tree of this basin, so it is an element present in the landscape of this region and the countries that make it up: Greece, Italy, Spain and those that make up the North African region as: Tunisia, Algeria, Morocco, and so on.

Currently, olive cultivation tends to spread to other regions of the world with similar climatic characteristics to those of the Mediterranean, although from the time of the conquest of America, began to spread to some of its colonies: Northern Mexico and now south western United States, also in Argentina there are news of its presence since the late eighteenth century. This intense current propagation responds to the goodness of its oil and its high economic value.

The flowers of the olive tree, from which the fruits later sprout are white with greenish tones and are arranged in the plant in the form of clusters with calyx in the dome and corolla of four petals. They have two stamens. In the pollination process the pollen is transferred from the stamens of a flower to the pollen of the same flower or other flowers of the same tree, or neighbouring plants.

Weeks before flowering water and nutrients are reduced, which causes the number of flowers per inflorescence to decrease.

The olive tree has several subspecies distributed in various areas of the planet, from South Africa, China, Macaronesia, and of course the Mediterranean Basin.

A very important aspect for olive growers is the alternation of good and bad harvests, which takes place as a natural process of the plants, in which their hormones intervene, and without having until now a complete understanding of the problem and how to avoid or modify it. Some varieties suffer this phenomenon more than others.

The fruit of the olive tree, olive, as it is commonly known, consists of the following parts: peduncle, mesocarp, epicarp (fleshy part), endocarp (stone) and embryo (seed). In the course of its growth it changes tonality (like many fruits of other plants): from green to yellowish green, later, when the harvest begins is when tonalities or purple dots appear, and at the end it takes a bluish black coloration.

The sowing or propagation of the olive trees can be carried out in very different ways, although generally it is carried out by seeds or by means of stakes. The latter is the technique most used to allow, through cloned propagation, plants generically the same as the predecessors, maintaining the purity of the species.

As a general rule, stems or stakes four or five years old are used, which when buried, over time, will produce roots and new stems, that is, a new plant. The rooting can also be done in nurseries, and then be transplanted, which improves the efficiency of sowing to have many new plants concentrated and facilitate the work of care. In short, there are many ways to carry out the propagation given the goodness of the olive trees in this sense.

Due to its ease of propagation is that currently have olive groves in many regions of the world, such as Argentina, Chile, Uruguay, Brazil, California, South Africa, Oceania, the Middle East, among others, but prevails as the centre and most important area of the Mediterranean Basin, as well as an ancestral culture in its people related to this crop.

To choose the variety of olive tree to plant it is necessary to take into account the composition of the soil, the climate: average temperature and maximums in summer and minimums in winter, the average time of sunlight, rainfall, humidity, etc.

For large areas it is preferable to plant two or more varieties of different ripening time in order to achieve a staggered harvest, which extends and improves the yield of production, thus lowering costs.

The average population of trees per hectare depends on the factors mentioned above, although it is generally between 200 and 300, so that the plants do not have to compete with each other for water, light and nutrients, and begins a rapid production with high yield. In adulthood, trees should not share shade with each other, as this way they appropriate less sunlight, which is necessary for photosynthesis.

The increase in the population of plants per hectare can be carried out until the previous figures are multiplied, as long as there are adequate varieties and an efficient agricultural technology, including irrigation and mechanization of the harvest, with which the results can be very favourable as it has been appreciated in some crops in Spain.

Under appropriate conditions, the plantations begin to bear fruit between the 3rd and 5th year, and in the 9th year they generally do not reach an optimum level of production. Trees can continue to be exploited for dozens of years, maintaining adequate pruning as they age.

Although there are hundreds of varieties of olive trees the most used to produce oil are: Picual (Jaén), Picudo (Baena), Hojiblanca (Lucena), Arbequina (Lérida), Empeltre, (Aragón), among others.

The Picual is the most cultivated in Spain, it is easy to propagate by stakes, early and high production, self-fertile, machinable and resistant to cold. Its oil is of great quality with a high lipid profile of oleic acid and little oxidable, what avoids, or delays the rancidity; its flavour is very strong and intense.

Weevil: It is widely grown. It is characterized by a precocious, high and alternating production. It also has a high pollinating and rooting capacity. The fruit is widely used as a table olive.

Hojiblanca: Clear leaves, which give rise to its name, has a discrete but sufficient rooting capacity and medium to late flowering. The fruits are strongly attached to the branches thus making it difficult to harvest by mechanical means. Also the pulp and the stone are very adhered to each other, which makes their separation more difficult. On the other hand, it should be noted that this variety is very resistant to drought and cold, as well as not very demanding with soils.

Arbequina: It has a high rooting capacity, it is not very vigorous, which facilitates its intensive production; it is of early production and presents high productivity. The fruit is little retained in the trees, but is small, which makes mechanized harvesting difficult.

Empeltre: Shows low rooting capacity which makes it advisable to propagate by grafting. Its entrance in production is late, it presents early flowering and high and constant production. The fruits are weakly retained, which facilitates the mechanization of the harvest. It has a high content of very aromatic oil with a light yellow tone and a sweet taste. It is resistant to drought and can be cultivated in relatively arid soils.

Pests and diseases.

Olive trees, given their longevity, are attacked by various fungi, bacteria and phytophages, within which they are found mainly:

Fungi: *Spilocae oliaginea, Fomitoporia punctata, Capnodium, elaeophilum*, among others.

Bacteria: *Pseudomonas savastanoi, Xylellafastidiosa*, among others.

Phytophages: *Bactrocera oleae, Prays oleae, Oaissetia oleae*, among others.

In a general sense, to fight pests, diseases and weeds in olive crops, different means are used: biological, biotechnological, physical, typical of classic farming methods, and chemical. In the application of control measures, these are carried out when the levels exceed a critical threshold, which determines the implementation of the means of choice, always trying to use the most appropriate.

In the case of the use of chemical products, it is necessary that these are adjusted to the characteristics of the diseases and pests to be combated, and that they entail the least risk for human beings, animals, beneficial insects such as bees, or others related to the biological control of pests.

The pests and diseases that attack olives, as well as the blows and fractures they may suffer during cultivation, and mainly during the harvest, considerably affect the quality of the olive oil obtained and its classification as virgin suitable for consumption.

Xylella fastidiosa

In the last four years the alarms have sounded intermittently in the olive oil sector of the Mediterranean Basin, where the largest production of olive oil in the world is centred, motivated by the bacterium known as *Xylella fastidiosa*, baptised for its ravages as *"the Ebola of the olive trees"*. This dangerous bacterium caused intense damage and substantial economic losses in the Italian region of Apulia, where the outbreak was detected in October 2013 and where from then on it is blamed with the death of more than one million trees.

It has been reported that after the affectation of the aforementioned Italian olive groves, *Xylella fas*tidiosa passed to Corsica and Provence (France), and at the end of 2016 to the Balearic Islands, Spain, and more recently, some fruit trees sensitive to the bacteria were detected in the Community of Valencia and even one case in the Community of Madrid itself, although so far no appreciable damage has been reported in the Iberian Peninsula.

This bacterium is lethal for olive trees, so they have been retaken and all necessary measures are being taken so that they do not reach the main regions of Hispanic olive groves, especially Andalusia.

The bacterium does not affect people or animals, but it is lethal to olive trees and other related woody trees, obstructing the circulation of raw sap through the plant. Its transmission is aided by insects. The final result on an olive tree can be the drought of leaves and branches that lead later to the death of the plant. It is difficult to study its effect in depth, as it can be accompanied by other pathogenic fungi.

The optimal growth of the bacterium is at temperatures between 26-28 °C and acts on the xylem of the plant. The insects that transmit it are generally sage suckers. Below 10 °C and above 34 °C their activity is limited.

In its control and treatment it is used the pruning of infected branches, thermal methods by means of jets of hot water, chemical control of the transmitting insects, as well as bio-control by means of other species of micro-organisms. At the end of the chapter, this delicate subject is explored in greater depth.

Composition of olives:

The fruit of the olive tree: the olive possesses a high quantity of vital nutrients for the human organism what constitutes one of the basic aspects that gives value to the olive oil, besides the high proportion of unsaturated fats, mainly monounsaturated -**omega** 9-, like the oleic acid that it contains. Many of these components remain in virgin or unrefined olive oil, also providing stability and durability to the product.

Broadly speaking, the percentage composition of the different components of the olives is as follows:

Fat: 16,3 %.

  Saturated: 2,03 %
  Monounsaturated: 11,3 %
  Polyunsaturated: 3,03

Carbohydrates: 4,4 %
Sugars: 0,55 %

Fibre: 3,85 %

Protein: 1,03 %.

Vitamins, at a lower percentage: A, B, E, K.

A (Retinol), β- carotene, Thiamine (B1), Riboflavin: (B2), Niacin: (B3), B6, Folic acid: (B9), E (tocopherols) and K.

Minerals: Na, K, Fe, P, Ca and Mg.

Water: 50

A more detailed study on the average composition of the fruits, and of olive oil in particular, will be dealt with later, as this is what gives importance and usefulness to this oil and the other marketable products of the olive tree, such as olives.

Main olive producing countries in the world.

As it is to suppose, the main producers of olives at world-wide level belong to the Basin of the Mediterranean, although it is to expect that in next times this distribution begins to change of appreciable way, attending to the intense proliferation that this culture is having in the last years in different regions of the world.

In 2011 the top ten olive producing countries were:

Main olive producing countries in the world (year 2011).

Rank	Country	Production (TM)	Area cult. (ha)	Rend. kg/ha
	World	20 545.4	10 057.6	2 048.8
1	Spain	7 820,1	2 503,7	3 123,4
2	Italy	3 182,2	1 144,4	2 780,6
3	Greece	2 000,0	850,0	2 352,9
4	Turkey	1 750,0	786,3	2 225,6
5	Morocco	1 415,9	900,7	1 571,9
6	Syria	1 095,1	684,5	1 599,8
7	Algeria	610,8	311,9	1 958,1
8	Tunisia	562.0	1780.0	315.7
9	Egypt	459, 0	52.7	8 727.3
10	Portugal	443, 0	343,2	1 293,1

Xylella Fastidiosa

CHAPTER IV

Complementary Reading

Xylella fastidiosa

As if olive oil, and more than this, olive trees, did not have more than enough with the fierce competition and all-out war that is waged in the sector of edible vegetable oils, where sometimes nothing is what it seems, now a new and lethal enemy emerges, a very small species of tiny dimensions, of the order of microns, in order to "spoil the party" infecting as much olive tree can find in its path, and not only these, but also other fruit trees and woody trees, which is his specialty. We are referring to a bacterium that annoys, bothers and can cause great damage to the European olive-growing sector, and in general to that of the Mediterranean Basin, and it is not that it can cause, but that it has already caused in Italy with the affectation of more than 500 000 olive trees. The bacterium in question is the Xylella *fastidiosa*, and it will be dealt with next.

The previous number of olive trees affected by *Xylella* may seem exaggerated, but it is not so, and it is not that the owners of the olive groves of Apulia in Italy want to lament or cry more than they should, it is that for some small producers these plants are like their children, for the purposes of the care and love with which they treat them, but more than this, like their parents, for they have been born and grown by their side, enjoying their shadow, finding beauty in those sometimes ghostly trees that inhabit the surroundings on their walks during the sunsets, but they have lived by their side the adventures and misadventures of good and bad times, happiness and tragedy, and more than this, life and death, birth and perish, living and dying.

There would be no words to describe the lament of the small owners of olive groves of remote times, who day by day contemplated their centenary olive trees awaiting harvest, some years good and others bad, sometimes with the obtaining of a product of extreme quality and others with the need to send it to the great refineries, because for certain reasons: biological, climatic, generally alien to their dedicated and sacrificed work, they cry inside, with twitchy fists, the impotence of not being able to do anything before external factors alien to their daily actions.

Xylella fastidiosa was not born, or was discovered in olive groves, not even in the Mediterranean Basin, but in fields of vineyards in Northern California, and not now, but in the eighth decade of the nineteenth century. It was there that Newton B. Pierce found it, commissioned by the U.S. Department of Agriculture in 1889, and although his studies did not reach conclusive results, his discovery is related to him, which still causes damage of more than 100 million dollars to Californian winegrowers.

Xylella fastidiosa is a phytopathogenic bacterium of the genus Proteobacteria that owes its characterization to Wells and his collaborators in a relatively recent time, in 1987, as one of the causes of its name: "*fastidiosa*" was the difficult thing to study and to characterize, as much as *"fastidia"* in the cultures, especially fruit trees of great economic importance, like the vineyards, peach trees, plums, almond trees, among others, to more than as it is studied in this case, to the olive trees, and many more woody plants are fruit trees or not.

Regardless of the damage it causes in America, in 2013 it appeared surprisingly in Italy, in the region of Apulia, in the south of the peninsula, where it destroyed more than 20,000 ha of crop, with the loss of hundreds of thousands of trees, but the bacterium has allies, powerful allies: the insects that transport it from one place to another with their own speed of movement, so that three years later, in 2016, it began its walk through the Balearic Islands, shortly after landed in Valencia and infected trees have already been found in the centre of the Iberian Peninsula, Madrid, although the damage caused is not remarkable and nothing similar to what happened in Italy.

But before this situation, it is normal that fear takes hold of the Hispanic olive growers, especially since this is one of the most important lines of its economy and provides employment to hundreds of thousands of people, accompanied by profits for the country of the order of billions of euros, but in addition, Olive trees have been part of Hispanic history and culture since time immemorial, and they are like bulls to the Spanish, as in plants, and perhaps even more so, because there are many more olive trees than bulls, and few consume bulls, but if many olives and olive oil, in addition to living from them. It is something like part of the Hispanic pride, also the Italian, Greek, Tunisian and every region that has taken the olive trees as a source of subsistence.

As an edible vegetable oil, olive oil is considered the best of all, the most complete in terms of beneficial effects for health and for the prevention of cardiovascular diseases, but it is also bow ship, the centre of one of the most important diets known: the Mediterranean, considered by many to be the most complete of all diets. And this is endangering its flagship, the one that guides it through the immense nutritional abysses, and we do not see that there can be a *Mediterranean diet* without olive oil.

It is difficult to understand how a small microorganism of a few microns in length can cause so much damage and deterioration to plants, but it does, and in what way, and to what extent.

But how does this bacterium work? It does so in a very simple but effective way and also at the same microscopic scale: it plugs the vessels through which the sap that transports the nutrients to the plant circulates, so that the cells of these in branches and leaves, when they do not receive nutrients they die and in the end this happens with the whole plant.

But there does not end the damage of the bacterium, this can happen to insects of the *cicadellic* genus, that is what is considered, when they feed on the xylem of the plant and from there travel comfortably to others, where it lodges immediately starting again the destructive cycle, and as one insect can damage several plants, occurs as a chain reaction, as in nuclear weapons with contagious and devastating effect.

Luckily, the damage in the Balearic Islands was mainly to the almond trees, but what would happen if they infect the olive groves of Jaén in Andalusia and ultimately in the South of Spain, the main producer of olive oil in the world, and more than main, the one that produces more than half of the entire world volume, as well as being the main consumer and exporter.

"If you see your neighbour's beards burning, soak your own", says the old proverb our parents and grandparents used to tell us, and this seems to be happening. Everyone fears this dangerous plague, this small bacterium a few microns in size - how complex nature is! and just as the atoms, even smaller, are the bricks with which the Universe is built, these small microorganisms can, in an adverse effect destroy a whole productive culture, although we hope that the current technological means manage to stop the advance of this bacterium, because it is not the first time that human beings face the action of our small companions, the majority of times benign, but also malignant as in this case, although for the bacteria these only do what their cellular structure and physiology dictates to them: To be born, to develop, to reproduce, and to die in any of the possible circumstances, and well that they execute their function.

In order to have a notion of the damage that can suffer the sector of olive growers by this and other problems, suffice it to say that from year to year the production of vegetable oils of different oilseed species increases in the world, however, the olive trees, although their planting has already spread to almost every corner of the planet, remains stable, For example, for the 2017-2018 season a total production of only 2,894 MT is expected, nothing comparable to the more than 60 million of palm oil that is considered to be its production volume in the same year.

In terms of the possible incidence of *Xylella*, world olive oil production in the 2013-2014 season was 3.252 MT, of which Spain produced 1.782 MT (72.5 %) and Italy, the second world producer 0.464 MT (14.3 %), but the following year a sharp decrease was manifested: world production fell to 2.458 MT, with a decrease of 0.794 MT. In that season Italy's production was only 0.222 MT with a decrease of 0.242 MT, more than 50 % more likely due to the incidence of bacterial plague.

But how to fight the pest?

Xylella fastidiosa lives in the plant xylem and there are insects that feed on it, such as the cicada, for example, which can spread it and become a vector transmitting the pest to other plants uninterruptedly, with a rapid and uncontrollable spread of the disease.

There are more than 300 species of trees sensitive to the attack of the *Xylella fastidiosa*, of which there are several subspecies like the **pauca** that was the one that infected and destroyed the olive trees of Apulia in Italy.

As an initial control measure, the transit of tree species that could be attacked by *Xylella* within the European Union is being avoided, although they are from seeds considered to have had nothing to do with propagation.

The formation of buffer zones outside the infected crops is also oriented to avoid propagation. Sometimes these can reach distances of more than a dozen km.

Around the infected plants, in addition to eradicating these in 100 m around, all plants possible to be attacked are eliminated, whether olive trees or not.

The area subject to emergency measures by the European authorities in Italy was more than half a million hectares. In the province of Lecce alone, in the Italian region of Apulia, this was more than 300 000 ha with more than 10 million olive trees quarantined, many of which, around 3 million, are around 100 years old, although they were found to be in good phytosanitary condition, but in any case, as the old saying goes, "it is better to be careful than to be sorry".

The subspecies that affected the Balearic Islands are the multiple, *fastidiosa* and *pauca*, which also attack the almond tree, lavender, cherry trees, myrtle, rosemary and vine, among others. In Alicante, two outbreaks affecting almond trees were detected, attributed to the multiple.

The possible vectors that can reach Spain are gardening trees from America, as well as coffee and fruit trees, as well as land from other infected regions. It is also necessary to take into account the material introduced by the movement of people or by travelling insects.

In the region of Apulia, in Italy, *Xylella* spread preferably through a common insect in olive trees known as *"foamer"* (*Philaenus spumarius*), this is a polyphagous entity abundant in olive trees.

There are no known practical means for the cure of the plant once infected. Systematic pruning, fertilization, and irrigation can help prevent disease. There is insufficient evidence that fungi and other pathogenic diseases affecting olive trees can facilitate infection with *Xilella*, although some have assumed, or believe otherwise, given the advanced age of many olive trees and their sub-optimal phytosanitary status, or lack of systematic controls.

The only way to fight the disease is against the transmission vectors, and the eradication of diseased plants, although it has been found that the severe pruning of some diseased parts has made possible the birth of healthy shoots. But this is not conclusive, and keeping infected plants alive may facilitate the spread to other healthy ones.

In vector control it is necessary to carry out phytosanitary treatments ranging from the elimination of weeds where the life cycle of insects is completed, to the use of phytosanitary products before the elimination of infected plants, following established agricultural and sanitary standards.

It is necessary that the felling around the infected plants is clean and total, because there can be focal points of infection.

Xylella fastidiosa is evolving and new subspecies may emerge, in addition to existing ones, with the problems associated with its control. Within its characteristics it is necessary to emphasize that it is a Gram negative bacteria with cell wall. It has vertical mobility within the xylem of the plant, as well as adhesion, but not flagella.

As it was expressed previously, the active subspecies known until now of Xylellas are: *fastidiosa, multiple, pauca and sandyi* that attack different types of trees and shrubs.

For example, in 2012 infected plants were found in France, and two years later, in the Netherlands in 2014, coffee plants imported from Costa Rica infected with *Xylella fastidiosa* were detected.

Therefore, the control measures implemented in the fight against *Xylella fastidiosa* consist of:

CONTROL MEASURES

1. **Exclusion**. Prevent other organisms from entering the area where the bacteria are present. Border controls regulate the entry of trees from outside the unaffected regions.

2. **Eradication**. Eliminate any plant showing symptoms of the disease. This measure only shows limited effectiveness in the first moments of detection of the pest and with a certain number of trees. Correspondingly, it is necessary to eliminate the asymptomatic trees that surround the place of infection, which is

a necessary prevention measure, but at the same time brings with it a high economic cost, as well as problems, litigation and the application of compensation policies with the affected growers.

3. **Escape**. This avoids contact with the inoculum of infection, and it is necessary to produce mother plants free of *Xyllela fastidosa* in greenhouses covered with nets that avoid the presence of vectors.

4. **Resistance**. This measure tends to obtain plants resistant to the disease, which is ultimately the most appropriate method, but is accompanied by a long and costly research process, for example, have been achieved hybrids of mandarin and orange immune to the attack of this bacterium. Transgenic seeds have also been developed but this clashes with the legislation of many countries, including the European Union itself.

5. **Cultural and chemical**, to decrease plant inoculum and exterminate disease vectors, including: systematic pruning and elimination of nearby weeds, treatments with chemical agents - insecticides, herbicides, etc. -Finally, to eliminate the vegetal cover and to control zones that can serve as lodging and survival of the vectors.

THE OIL OF THE OLIVE TREES

CHAPTER IV

The Oil of the Olive Trees

Olive oil

Olive oil, name that said in a table of passionate people to the kitchen represents something like the universal panacea, or the gold for the seekers of fortune, such are considered the kindness of this type of oil originating in the Basin of the Mediterranean and rich, very rich in fatty acids of the type omega 9, like the oleic one, and also antioxidants, polyphenols, vitamins, and minerals, among other secondary components considered beneficial for the health.

Olive oil is a vegetable oil that is extracted by pressing and then filtering, from the fruit of the olive tree, *olive*, so it differs in this way from seeds such as sunflower, rapeseed, soybeans or corn, which are then subjected to a refining process. This is possible because the organoleptic qualities of virgin olive oil allow it and more than a fifth of the whole fruit is oil, as well as around 30 % of its pulp.

Olive oil is mainly used in cooking to prepare and prepare food, unlike other oils that have various uses in industry, food, bio-fuels, cosmetics, etc., which is due above all to its high quality for human consumption and the high price it reaches in the market, 4 or 5 times higher than others, given the high demand and agricultural production costs in a type of harvest that is not very machinable, with a yield per hectare lower than other common oils.

The extraction of the juice from the olives is relatively simple, and in immemorial times it was carried out in rustic stone mills:

mills, which can still be found as museum objects and also in some traditional companies.

The process of extracting olive oil is so simple that it can be prepared even in houses, with very limited resources, simply counting on a simple grinding machine and a press operated with a hydraulic jack of automobiles. Of course, at present this practice is neither recommended nor necessary, especially in cities and in areas far from crops, for many practical and economic reasons that are obvious to the naked eye.

The extraction of the oil is done taking fresh olives as a starting point in their initial ripening phase (6 to 8 months), which coincides with their optimum yield stage. These, taken to the factory and cleaned of leaves, remains of branches, etc.. are washed, then ground and finally filtered the semi-solid cake, yielding an oil from green to dark green (virgin olive oil, or extra virgin), which shows a characteristic flavour, acidity and bitterness, due to non-oily substances that accompany the juice, which far from constituting undesirable compounds, increase the value of this oil, as they are rich in vitamins and natural antioxidants such as carotene, niacin, tocopherols and polyphenols, among others.

Unlike other oils, the crude extract constitutes the basic product or star of the olive oil industry, it is of great stability due to the presence of natural antioxidants and contains many bioactive substances valued as benign and beneficial for health.

Depending on the quality of this first oil determined by experts through traditional tasting: smell, colour, flavour, etc., it is taken to pack directly as virgin olive if it meets the required indicators, or glaring if its quality is not suitable for immediate consumption, so this last fraction is sent to the refining process.

The glaring fraction, once passed to the refining process in union with other products derived from the manufacturing process, and with the use of organic solvents, loses the virgin term and the final result will be a refined olive oil similar to other commonly used vegetable oils, whether from seeds or not.

The oils obtained by refining are transparent, have a clear appearance: green or yellowish green, even slightly yellow, which is classified according to acidity: 0.1 and 0.4 degrees, respectively. They have lost a large part of the initial products beneficial to health, so its quality is lower than that of virgin olive, although its price is still higher, between two or three times that of other oils, such as sunflower, to take it as a reference.

Also from the stone of the fruit, or seed of the olive an oil of inferior quality is extracted that is named of *marc*, that also is commercialized.

Olive oil is susceptible to alterations in its chemical composition due to the effect of the radiation of greater energy from sunlight (**UV**), which is why glass or coloured plastics are used in its commercial packaging. Generally, green glass packaging is the highest quality and sometimes corresponds to factories that have on their labels many years old (some of them dating back to previous centuries), designations of origin, cultivation area, etc..

The production of olive oil is closely related to the ancestral norms and customs of its producers, and is not totally limited to technical cultivation indicators such as those of other oleaginous plants.

In the last months of the year, the time when the fruits yield the most oil, the harvesting of the olives begins, previously carried out by means of the milking and beating system, the first directly with the hands and the second consisting of hitting and forming strongly with a stick the branches loaded with olives, some relatively high, that fall on a canvas or similar mantle that covers the ground, so that they are not infected with soil or more impurities. This process is preceded by a good cleaning of weeds and weeds, as well as the smoothing of the ground to facilitate the laying of the covers on a completely smooth surface.

Formerly leaves, branches, wood chips and other impurities were separated manually in the same field, work usually done by temporary women, but now is executed in factories using mechanical means.

The *beating system* has been modified in most of the cases by means of the use of mechanical vibrators adjusted to long rods, with which the branches are shaken alleviating the human effort and increasing the yield, also the farms of greater extension, when the type of fruit and the conditions of culture facilitate it, they go to agricultural machineries that surround the plant, they adjust it like with a tweezers, and they shake it strongly making even the harvest more efficient. In any case, this method must be completed with the beating method to complete the separation of the fruits that remain without detaching.

The fruits are transferred immediately to the oil mill or factory, because they deteriorate quickly altering their composition, taste, aroma, etc., that is to say, their quality and that of the oil to extract.

In spite of the mechanical shocks caused by the means of harvesting used, it is necessary for the fruit to suffer as little damage as possible, especially breakage, in order to prevent the entry of fungi that can affect the organoleptic characteristics of the oil and, in general, the performance of the factory. For this reason, it is necessary to differentiate between olives that have fallen to the ground and have received some kind of blow, or even trauma to their structure, from those taken directly from the tree by means of the ancestral milking method.

In general, the harvesting system should be applied that harms the fruit the least. In any case, manual milking is the most appropriate method to obtain a best oil, of course this makes the process and the product, already relatively expensive.

Once in the factory, the olives are cleaned and sieved to eliminate leaves, stems, soil, etc. and they are washed with water at room temperature, which eliminates the dust adhered to the olives for months outdoors, as well as the remains of pesticides, if the olive trees have been treated with these products. It is recommended that the fruit be processed within 24 hours of harvesting.

In the milling process the tissues of the fruit are broken so that the oil can be expelled from its cells, a process in which metallic hammers are generally used spinning at high speed. The obtained paste is beaten slowly during a determined time to facilitate the extraction of the oil of the cells, question that generally is made to a temperature average of 25-30 °C so that the viscosity of the oil does not increase while the oily mass is formed.

The beating time must not be excessively long, so as not to affect the loss or transformation of products vital to the quality of virgin oil, such as tocopherols and polyphenols, and to achieve virgin olive oil with the best possible colour, flavour and aroma, as well as the other desired organoleptic factors.

In order to separate the oil from the paste, which contains water, and the rest of the solid ground components of the fruit, including the crushed seeds (marc), this is subjected to a discontinuous extraction process by pressing, a traditional method almost in disuse, or even better, by means of a continuous process of centrifugation, in which the paste passes through centrifuges that separate, by difference in density, the oil from the water and the marc. This process is continuous and very efficient.

Within the different variants of the method of centrifugation, in some of these water is added before beating the mass, but it excels more than this known as two-phase, that of directly centrifuging the original oily mixture without adding water, in which solid and liquid waste come out together, being more complex the residual treatment because it contains three quarters of water, so to dehydrate it is used at slightly higher temperatures, but on the contrary, much less water and energy is used making the process more efficient.

The oil that comes out of the separation of the rest of the components of the cake (virgin olive, or extra virgin olive), with a yield close to 20% of the fruit, is feasible to use without suffering appreciable alterations at least during a year.

The remaining cake from the separation process still contains appreciable quantities of oil and once the accompanying liquid is separated it is known as marc, and its oil is extracted using organic solvents as in the rest of the processes of refining seed oils. Crude olive oils that did not pass the compulsory tasting tests and present inadequate levels of acids or other undesirable indicators of flavour and odour (lampantes) are also added to this process.

The refining process, in the case of olive oil, consists of the following steps:

-**Winterization**: This is the process followed in the refining of vegetable oils in order to eliminate, by means of precipitation due to a drop in temperature, poorly soluble materials such as waxes, molecules derived from fatty acids higher than stearic acid, etc. For this, the oils are cooled to eliminate the glycerides with the highest boiling temperature (stearates, waxes, sterols, etc.) generally associated with saturated carbon chains, which solidify to be separated by subsequent filtration. Generally, this process, as it is carried out in the winter months, does not require cooling, but when there is a great demand, conventional means of refrigeration are used. In this stage it is necessary to let the product rest at low temperatures (5 °C) for approximately 24 hours (1 day).

-**De-resinification**, or de-gomification. It is the process in which the mucilages and gums are separated, including the lecithins present in the mixture.

-**Decolouration**. It is carried out by means of activated carbon or another adsorbent material, and the substances responsible for intense colours, such as very dark green and brown, corresponding to oxidised products, are eliminated.

-**Neutralization**: With it is possible to reduce the degree of acidity or pH by means of the treatment with alkaline solutions of sodium hydroxide, in a process similar to saponification, producing soaps of the free acids insoluble in the oil, for its relatively polar character.

-**Deodorization**. This process is carried out by means of superheated vacuum water, using temperatures between 160-180 °C and in it the most volatile substances are expelled, such as aldehydes and ketones, which cause intense undesirable odours that can significantly alter the quality of the product.

Some of these latest technological processes are carried out in modern refineries, as traditional mills or mills do not have adequate equipment and technical means to carry them out efficiently.

Through these processes, the olive oil (now *refined)* loses most of the biologically active products and health benefits present in virgin olive oil, so it is generally mixed with a certain amount of it in a certain proportion (10 to 20 %), which makes it a *rectified oil*, which does not contain the active components in high proportions, but in relatively adequate quantities, so they are in this sense superior to traditional refined seed oils.

Due to the above, the scheme for obtaining olive oil loses its apparent simplicity and a series of different oils appear on the market in terms of their composition, fundamentally in terms of the unsaponifiable materials that accompany the glycerides, which gives rise to different types and qualities of oils.

In view of what has been said about the diversity of oils and in order to protect the original products from possible, perhaps frequent, adulterations that olive oils may suffer, the European Union (**UE**) and other international bodies such as the International Olive Oil Council (**COI**) have established classification standards of strict control which, in the case of the **UE**, are detailed below:

-**Extra virgin**: Obtained exclusively from olives in good condition by means of mechanical procedures, with an excellent smell and flavour according to tasting by specialised professionals, with a degree of acidity of less than 0.8° and with an average of 0 in fruitiness and sweetness. It cannot have a concentration greater than 20 meq/L of active oxygen (peroxide index), which is responsible for the rancid or defective taste of the oils.

Virgin: Obtained according to the same requirements as extra virgin olive oil, although with a higher degree of acidity, between 0.8 to 2°, with indicators of defects lower than 3.5, not perceptible, and fruity 0. Its flavour is milder and contains a lower concentration of antioxidants than extra virgin.

Olive oil: Mixture in appropriate proportions of virgin olive oil, and refined obtained from the oils marked defective to have not reached the basic indicators of quality of virgin olive, and the remnant not extracted from the cakes during the process of virgin olive production.

In the process of obtaining the generically named olive oil, chemical methods of purification are used, so that its quality is inferior to that of the virgins with regard to the proportion of not saponifiable biologically active components. The degree of acidity in this oil cannot be greater than 1. They are also shown in markets as *mild* or *intense* olive oil, an issue not included in EU standards, and which in our opinion may confuse the consumer.

Marc oil: It consists of a mixture of refined olive marc oil extracted from the cakes where the virgin oil has been separated, and then mixed with part of it, without the acidity of the final product being greater than 1.5°.

It should be noted that in addition to the oils listed above, is the so-called *glaring*, as it was previously called for its use in lamps for lighting. This is a crude oil not recommended for human consumption because it contains undesirable substances, especially organoleptic, due to fruit components not optimal for oil production, but suitable for refining, and then mixed with a proportion of virgin oil of 10-20 %, is sold under the generic olive oil.

The basic characteristics of glaring olive oil are: acidity greater than 2°, tasting defects greater than 2,5, and the median fruitiness equal to 0. In general, it is very acidic, with flavours and odours if not unpleasant, far from the original virgin olive oils.

This diversity of genres makes it complex for the consumer to buy olive oil according to their requirements, as the general public is unaware of the indicators by which this classification is governed, which is added to the wide variety of oils of different types that are marketed, this question, which in the opinion of the author, should be cause for reflection by the agencies involved, because people should and want to consume the right oils according to the use to be given, and among so much confusion of types and names, virtually no one is right with what it seeks, and can not be described as profane, as it is almost the majority of consumers.

Finally, it should be pointed out that this European regulation is not equivalent to those used in other countries, such as, for example, the United States, one of the main consumers of olive oil in the world, and which bases its assessments on acidity and the absence of defects in taste and smell, according to the regulations of the United States Department of Agriculture (**USDA**) of 1948, which defines the following types or grades according to:

-**U.S. Grade A or U.S. Nancy**: Olive oil with no more than 1,4° acidity, free from organoleptic defects.

-**U.S. Grade B or U.S. Choice**: Olive oil with a maximum of

2,5° acidity and a moderate amount of defects.

-U.S. Grade C or U.S. Standard: Olive oil with no more than 3° and some organoleptic defects.

-U.S. Grade D or U.S. Substandard: Olive oil with an acidity index higher than 3°, accompanied by defects that can be appreciated, but within the quality frames required by the **USDA**.

Composition and properties of olive oil

The properties and use of olive oil, like those of any vegetable oil, are determined, in the first place, by its composition or fatty acid profile, although it is appropriate to point out that a significant part of the goodness of olive oil derives from the secondary components which accompany it, and which represent less than 2% of it, but which are responsible for some of the most important properties of this oil.

With regard to the above, it should be noted that in olives, in addition to the triacylglycerides, there are many other substances in minority composition such as: Vitamins, tocopherols, sterols and polyphenols, among others, which, once extracted from the oil mills, are transferred to the oil mills and remain in them, guaranteeing their stability and durability, as long as it is virgin olive oil, since when it undergoes the refining process it occurs, as in other vegetable oils, that it loses the reference components, or its concentration decreases appreciably, and with it many of the beneficial properties of these.

It is for this reason that emphasis is placed on packaging and marketing virgin olive oil as a component and main element of the olive oil industry.

As far as the average lipid profile of olive oil is concerned, this is shown in the following table:

Fatty acid profile of virgin olive oil (g/100 g)*

AGS	AGM	AGP
C14:0 Myrístic —	C16:1 Palmitoleic 0,9	C18:2 Linoleic 10,5
C16:0 Palmític 11,5	C18:1 Oleic 68,8	C18:3 Linolenic 0,7
C18:0 Steáric 2,2		C20:4 Arachidonic -
TOTAL 13,7	TOTAL 69,7	TOTAL 11,2

*Moreiras y col. (1992).

The high proportion of oleic acid (68.8%) is the most important aspect that stands out in olive oil, to which is added the relatively low amount of saturated fatty acids, which together do not add more than 14%, prevailing palmitic acid (C16:0) in proportions greater than 10%. It is necessary to emphasize the presence, although in smaller quantity, of palmitoleic acid, which also presents a hydrocarbon chain of 16 carbon atoms, but with a double bond in position 7 counted from the carboxyl group (omega 7).

The presence of 11.25% of polyunsaturated fatty acids where linoleic acid prevails (two double bonds in the hydrocarbon chain) is less feasible to oxidation than linolenic acid (three double bonds), whose concentration is significantly lower. These acids are of singular importance for the human organism because it is incapable of synthesizing them and they must come from the diet. Linoleic acid is found in high quantities in many other vegetable oils such as sunflower, corn, peanut, etc., but not the linolenic (omega 3) that although needed by the human body in much lower amounts than its similar smaller this contribution of olive oil is to be thanked, since it is not very abundant in other oils except for linen and in a much smaller proportion in canola and soy.

Already this lipid composition, with fatty acids relatively stable to auto-oxidation, makes it possible for olive oil more satisfactory results than other oils in frying, well above the oils of sunflower, soybean and corn, among others.

Also in stability studies during prolonged storage, without proceeding to the heating, the olive oil behaves more stable than the previous ones.

As far as human health is concerned the studies show that responsible and controlled intake of olive oil favours blood circulation and has a positive effect on atherosclerotic damage and the reduction of low density lipoproteins (**LDL**), which are responsible for the transport of cholesterol to the arteries.

In addition to the lipid profile, if we go to the components that accompany virgin olive oil, the unsaponifiable fraction, a little less than 2%, we find that because it comes from a fruit has numerous components beneficial to health such as, among others: squalene, β-carotene, chlorophyll, tocopherols, sterols, and polyphenols, the latter of marked antioxidant action, evidenced by the stability of virgin olive oil once packaged and its marked resistance to deterioration and rancidity.

Seen in this way, virgin olive oil behaves in an intermediate way, as a vegetable oil to which the properties of the olives are added, so that the varieties of the latter affect the characteristics of the oil obtained, as well as the form of harvesting and processing, and of course the conditions and time of storage.

Virgin olive oil is shown to be relatively more acidic than refined seed oils, which responds to the presence of free fatty acids typical of olives, or formed during the harvesting and manufacturing processes, which has nothing to do with taste and other organoleptic aspects, although they are an element to be taken into account in their classification as *extra virgin or virgin*.

The unsaponifiable components of olive oil are not eliminated in the mechanical process of obtaining virgin olive oil, but to a large extent in that of refined oil, which establishes a clear dividing line between these types of oils, as if they did not share their own nature. In this sense, except for its lipid composition, refined olive oil is more similar to seed oil, especially to high oleic and canola oil, enriched by genetic selection in oleic acid.

Unlike other vegetable oils, the compounds that accompany virgin olive oil and are responsible for its organoleptic properties play a crucial role in determining its quality and classification. So in Spain and other Mediterranean countries, a relatively subjective factor such as tasting by a group of experts ultimately determines whether an oil goes to market, and the category that corresponds to it according to the regulated standards.

Later, in the text, we will go back again to mention and carry out the study of these secondary components which make up the minor or unsaponifiable fraction of olive oil, and which are of considerable importance.

The expert tasters, in addition to the colour and the physicochemical indicators determined in the oils by laboratory techniques, stop to evaluate a series of organoleptic aspects such as the following:

Positive: Fruity (green or ripe), bitter and spicy.

Negative: Borras, mould-moisture, fungi and yeasts, avinized-vinegary/acid-sour, metallic, rancid, among many others only susceptible to be valued by those who practice this profession.

Refined olive oils are not subjected to tasting procedures, their quality being controlled by means of physicochemical indicators in the laboratories, among which the concentration of free oxygen, degree of acidity, among others, show an important significance. The intensity of the taste of this oil will depend on the degree or concentration of virgin olive oil that has been added to improve its quality, which is generally between 10 and 20 % to be generically designated as mild or intense, respectively.

The tone and colouring of virgin olive oils depend to a large extent on the type of olive used in their manufacture, and the nuances of one variety of fruit may vary considerably from one to another, for example:

The picual: The most widespread in Spain, produces oils of green tones and slightly bitter fruity flavours.

The hojiblanca: Oils with golden tones and a mild flavour.

The arbequina: Aromatic oils, slightly bitter and spicy, green, especially at the beginning of harvest.

The empeltre, on the other hand, produces yellow and sweet oils, with very fruity aromas.

And so on with other types of olives, in a game, or endless rule, in which, as a rule, each olive variety produces a certain type of oil.

Uses of olive oil

Olive oil, due to its characteristics, has multiple uses, especially food, among which stand out:

It is a basic product in the Mediterranean diet and, of course, intensely used for nutritional purposes by the populations that have inhabited this basin since antiquity.

It is widely used and accepted in the preparation of salad dressings, to which it gives the characteristic aroma and flavour of this oil, also mixed with lemon (Mediterranean): vinaigrette.

For the preservation of food, especially canned fish such as sardines, tuna, mussels among others. It is also used for the preservation of vegetables, meat, cheese, etc..

To prepare oils flavoured, with rosemary, basil, garlic, lemon, etc., even smoked. In the case of garlic and rosemary, the beneficial properties of the oil are combined with the antioxidants, anti-inflammatories, antimicrobials, etc. of these plants.

The fact that olive oil has a pleasant flavour and a characteristic aroma makes it possible to eat it directly with bread, garlic, tomatoes, etc., which constituted (and is still used) an efficient traditional breakfast for sustenance in southern regions of Spain such as Andalusia and in other areas of the Mediterranean Basin.

Heated in fried foods is one of the oils that most maintains its natural properties and with a lower evaporation loss than others, especially seeds, due to its lipid profile rich in oleic acid and its natural antioxidants, although with the increase or repetition of heating, these are altered in their chemical composition, progressively diminishing their antioxidant properties It also has a high smoke point.

The industrial use of olive oil for non-food purposes is relatively limited due to its high price however it is also used in the cosmetics industry in creams and oils, and to produce certain types of soaps.

Within its beneficial aspects it can be pointed out that:

Olive oil does not contain cholesterol in the sterols that accompany it as secondary products.

Olive oil is rich in vitamins A, E and K.
It facilitates digestion.

The polyphenols present in appreciable quantity in olive oil prevent degenerative diseases and cellular aging.

It reduces the risk of cardiovascular diseases by positively affecting the main risk factors: it increases the levels of high-density lipoproteins (**HDL**), decreases the levels of low-density lipoproteins (**LDL**) and also of total cholesterol (**COLt**).

It is a basic component of the Mediterranean diet.

As far as conservation methods are concerned, especially for virgin oils, certain procedures must be followed, such as: protecting them from light, maintaining them at a normal and constant temperature without great variations, and not exposing the liquid to the air to avoid auto-oxidation and rancidity. In short, keep it hermetically covered, in the dark, packed in opaque containers and at normal temperature.

Other Fatty Acids Present in Olive Oil.

In addition to the main fatty acids contained in olive oil such as palmitic, oleic, palmitoleic, stearic and linoleic, other fatty acids such as myristic, margaric, heptadecenoic, linolenic, arachidic, eicosenoic, behenic and lignoseric may be present in a much smaller quantity, which will be dealt with briefly below.

Myristic acid: (C14:0). Tetradecanoic.

$CH_8(CH_2)_{12}COOH$

It is a saturated fatty acid, solid at room temperature, with a medium to long hydrocarbon chain, made up of 14 carbon atoms, including that of the functional group carboxyl. It is not very soluble in water, but in solvents of lower polarity...

Molecular mass (M): 228,4 g/mol
Density: 0.8622 g/cm³
Melting temperature: 54.4 °C
Solubility 1,07 mg/L

Its name comes from the nutmeg (*Myristica fragrans*), whose solid fat contains high amounts of this fatty acid (75%) in the form of triacylglyceride or trimyristine, as it is commonly called.

Its concentration close to 20% in coconut oil is considered a risk factor in cardiovascular diseases, due to its positive correlation with low density lipoproteins that transport cholesterol. In olive oil its concentration does not exceed 0.05%, so its action is insignificant.

Palmitoleic acid. (C16:9). Delta-9-cis-hexadecenic.

$CH_3(CH_2)_5CH=CH(CH_2)_7COOH.$

Although it was briefly mentioned at the beginning of the chapter, some of its characteristics will be evaluated.

Is in a liquid state at room temperature

M: 254,41 g/mol
Melting temperature: -0.1 °C
Density: 0.894 g/cm³

It is a monounsaturated fatty acid with a lower hydrocarbon chain than oleic acid: 16 carbon atoms. It is a component of human adipose tissue fats. It belongs to the omega 7 series (ω-7). Its role in cardiovascular diseases is not entirely clear, and in another sense, reference is made to cellular oxidation of the skin. It is present in minority concentration in some vegetable oils such as those of the olive itself by an average of 0.9%, although it can be found in a greater or lesser proportion depending on the nature of the olives, the region of cultivation, as well as the techniques used in this. Palmitoleic acid is found in greater proportion in avocado oil (7 %) a product so far little marketed, also in much smaller amounts in oils of maize, soybean and sunflower (about 0.1-0.2 %). It is also present in whale oil (9 %), lard and beef: (2,5-3,5 %) and 2 % in butter.

Margaric acid (C17:0). Heptadecanoic.

CH₃(CH₂)₁₅COOH.

Liquid state, at room temperature

M: 130,8 g/mol
Melting temperature: -7,5 °C
Boiling temperature: 223 °C

It is a rare fatty acid in vegetables due to its odd composition of carbon atoms. It can be found in very limited quantities in olive oil, generally in concentrations of less than 0,3 %. It is found in the milk fat of cattle and similar animals, and their derivatives such as butter and margarine. Its main use is in the leather tanning industry and as a polishing agent.

Heptadecenoic acid (C17:1)

It is a very rare fatty acid, when it appears in olive oil never exceeds a concentration of 0.3%.

Linolenic acid. (C18:3n3)

Cis,cis,cis-9,12,15-octadecatrienoic.

COOH(CH2)8-CH=CH-CH2-CH=CH-CH2-CH=CH-CH2-CH8

Liquid state, at room temperature

M: 278.43 g/mol
Melting temperature: -11,0 °C
Density: 0.914 g/cm³.

Like linoleic acid, it is an essential fatty acid that cannot be synthesized by the human organism, which therefore acquires it through food. It belongs to the series omega 3 (α) and 6 (γ), is present in some vegetable seed oils such as chi and flax (greater than 50%), and to a lesser extent, but still with some significance in the canola (10%) and soybean (7%) which gives some instability to oxidation reactions, which implies that in edible oils where it is present are added antioxidants of a certain power to prevent rancidity and oxidation. In olive oil it can be found in concentrations lower than 0.9%.

Arachidic acid (C20:0). Eicosanoic.

CH3(CH2)18COOH.

Solid state at room temperature

M: 312,53 g/mol
Melting temperature: 75,5 °C
Boiling Temperature: 328 °C
Density: 0.824 g/cm³.

It is found in a greater proportion in peanut oil (1.0-1.7 %). In olive oil it can appear in concentrations lower than 0,6 %. It is a saturated fatty acid.

Eicosenoic acid (C20:1n9). 11-Eicosenoic acid.

$CH_3(CH_2)_8 CH=CH(CH_2)_8 COOH$.

It is an omega 9 fatty acid, although it does not belong to the essential fatty acids and the body can do without it. It is generally found in fish oil. In olive oil it can appear in concentrations lower than 0.4 %.

Behenic acid (C22:0). Docosanoic.

$CH_3(CH_2)_{20}COOH$.

White solid, at room temperature

M: 340.58 g/mol
Melting temperature: 80,0 °C
Boiling temperature: 306,0 °C
Density: 0.893 g/cm³.

Long chain saturated fatty acid that may be associated with atherosclerotic damage is found in very small proportions in rapeseed oil and peanut oil. In olive oil it can appear in concentrations lower than 0.2%. It is mainly used in cosmetics due to its softening action, as well as in the surfactant industry.

Lignoceric acid. (C24:0). Tetracosanoic.

$CH_3(CH_2)_{22}COOH$

M:368.63 g/mol.

It is found in wood tar as well as peanut oil in proportions of 1.0-2 %. In olive oil it can appear in concentrations lower than 0.2%.

Main producers and exporters of olive oil

Olive oil is produced in a smaller volume than other oils such as African palm, soy, rapeseed, sunflower and maize germ. World production is centred on Spain, Italy and Greece, and well above all the former, which processes almost half of what is obtained at world level. Although it would seem that Italy, given its traditional commercial propaganda was the first, but this is not the case, because the Iberian country exports a considerable quantity of virgin olive oil in bulk to it (in tanker trucks), which later, the Italian country may pack as its own under its traditional brands or as part of its own export volume.

In the 2011-2012 campaign, world olive oil production was 3.321 MT of which Spain produced 1.615 MT, representing 48.6%, i.e. approximately half. This situation continues with a slight percentage decrease. This indicates the importance of the olive sector for the Iberian country.

The world consumption of olive oil in recent years is shown below:

World consumption of olive oil - progression (ODEPA and COI)

CAMPAIGN	CONSUMPTION (MT)
2002/3	2,7
2003/4	2,9
2004/5	2,9
2005/6	2,7
2006/7	2,8
2007/8	2,8
2008/9	2,8
2009/10	2,9
2010/11	3,1
2011/12	3,1
2012/13	3,0
2013/14	3,0

As can be seen, olive oil consumption in recent years has maintained a slightly increasing trend and has risen by around 10 % over the last 12 years, which gives an idea of the high esteem with which it is currently held, although the demand for this product is largely limited by existing production capacities, since oil production cannot be increased abruptly given the cultural characteristics of olive-growing.

The main consumers of olive oil in the world are the European

Union (**UE**) (84%) and the **UE** (10.7%). Spain and Italy are the main consumers within the **UE**.

In other regions of the world olive oil production tends to increase, and in Chile, for example, the arable area increased by 100 % in the last 10 years from 2 000 MT in 2002 to 15 000 MT in 2014. Argentina, on the other hand, produces more than 20,000 MT per year.

It is estimated that there are around 2 000 brands of olive oil in Spain.

World production of olive oil in the 2016-2017 campaign was 2,539 MT, much lower than that of the 2015-2016 campaign (3,176 MT), with a decrease of 0,637 MT. This fall is influenced by data from the European Union, the main production area, which fell from 2.324 to 1.747 MT. For the 2017-2018 season, a certain improvement is expected in the sense that a world production of 2,894 MT is achieved, while in the EU the same occurs and this reaches 2,717 MT (**COI data for November 2017**).

The EU currently produces more than two thirds of the world's olive oil. The year of highest production of this oil since 1991 was in the 2011-2012 campaign (3,331 MT), which coincided with the highest production in the EU (3,064 MT), which shows the incidence of the old continent in the production of olive oil on a global scale.

Below is a table with the variations of world olive oil production in recent years for the main producing countries.

Country*	2011/2012	2012/2013	2013/2014	2014/2015	2015/2016
SPAIN	1.615,00	618,20	1.781,15	842,20	1.401,60
ITALY	399,20	415,50	463,70	222,00	474,60
GREECE	294,60	357,90	132,00	300,00	320,00
TURKEY	191,00	195,00	135,00	160,00	143,00
SYRIA	198,00	175,00	180,00	105,00	110,00
MOROC.	120,00	100,00	130,00	120,00	130,00
TÚNEZ	182,00	220,00	70,00	340,00	140,00
PORT.	76,20	59,20	91,60	61,00	109,10
CHILE	21,50	15,00	15,00	18,50	16,50
JORD.	19,50	21,50	19,00	23,00	29,50
AUSTR.	15,50	9,50	13,50	19,50	20,00
ISRAEL	13,00	18,00	15,00	18,50	15,00
EEUU	4,00	4,00	12,00	5,00	5,00
CHINA	-	-	-	2,50	5,00

* Production in Thousands of tons.

Source: http://www.internationaloliveoil.org/

It can be noted, according to these data, that the 2014/2015 season was very bad for the main producers, especially for Spain and Italy, although there was a period of strong recovery in the following year, with a decline again as projected for the 2016/2017 season, which responds to climatic difficulties in the Mediterranean area, which have affected crop yields, as well as the variability of yield of olive trees by their own natural characteristics.

HIGH OLEIC
OILS

CHAPTER V

High Oleic Oils

Until the mid-twentieth century, olive oil occupied a discreet place in the universe of vegetable oils and fats in general, with strong competent monopolizing this industry, especially cotton and soybean oils, margarines and hydrogenated fats, etc., attending to the purposes for which they were used.

Lard (lard), margarine and hydrogenated fats, among others, occupied a prominent place, but this was when a turnaround occurred when studies were begun to determine the incidence of fats in the metabolism of the human body and essentially on the risk factors of cardiovascular disease (**CVD**). Some of these were carried out on a large scale and over a prolonged period of time. They showed that fats with a high concentration of saturated fatty acids such as palmitic acid were related to the elevation of indicators that characterized atherosclerotic damage such as cholesterol and low-density lipoproteins (**LDL**).

At the same time, in later years, the incidence of *trans* fatty acids on **CVD** was also verified. These acids are formed when oils are subjected to high temperatures in the refining processes and especially in the catalytic hydrogenation to produce saturated fats, which turned the industrial sector of fats that began to give preponderance to the use of unsaturated fats, such as olive oil, which went from occupying a modest place to become the preferred food preparation in kitchens.

The lipid profile of a group of fats compared to that of olive oil is presented below.

Comparative lipid profile of solid fats and olive oil (%)

Fat acid	Lard	Tallow	Margarine	Butter	Olive A
C.14:0	1,5	3,0	1,1	8,9	-
C:16:0	25,4	26,5	18,8	20,6	11,5
C:18:0	14,8	19,5	4,1	8,9	2,2
T. Sat.	41,7	49,0	23,0	38,4	13,7
C16:1	2,4	3,5	1,1	2,1	0,9
C18:1	38,5	40,0	29,2	22,1	68,8
T. MI.	40,9	43,5	30,3	24,2	69,7
C18:2	8,2	4,5	16,7	1,1	10,5
C18:3	0,7	-	1,6	1,2	0,7
T.PI.	8,9	4,5	18,3	3,3	11,2

Fuentes:

Lardo: Belitz y Grosch (1997).
Sebo vacuno: Moreiras y cols. (1992)
Margarina: Moreiras y cols. (1992)
Mantequilla: Moreiras y cols. (1992)
Aceite de oliva: Moreiras y cols. (1992)

As can be seen in the table above, the average concentration of saturated fats associated with atherosclerotic damage in olive oil is about twice as low as in margarine, about three times as high as in butter and lard, respectively, and about four times lower than in tallow. These highly significant differences are conclusive for the preferential use of olive oil over the reference solid fats in food processing.

In addition to the above, if monounsaturated fatty acids such as oleic acid (C18.1) and palmitoleic acid (C16:1) are referred to, the difference between olive oil and other fats is also highly significant, exceeding solid fats by more than 30 percentage units. If we take into account that clinical studies show that these fatty acids do not raise the levels of **COLt** and **LDL**, and rather increase that of lipoproteins that have a beneficial effect on the body such as high density (**HDL**), it is understandable why olive oil began to play an extraordinarily positive role as a nutrient element and began to be quoted as the best oil when preparing food.

But even this oil had numerous competitors, not in the sector of solid fats, but in that of the vegetable oils themselves, as countless of these had significant amounts of unsaturated fats with beneficial and protective action on the LCA, such as sunflower, soybean, corn, etc.. However, luck was on the side of our protagonist when, with the passing of time, it was proved that these oils were not very stable when they were subjected to heating processes, oxidizing themselves easily with the appearance of intermediate free radicals, and in the end highly dangerous compounds for the body as peroxides, for its negative impact on cellular metabolism.

The lipid profile of some vegetable seed oils compared to that of olive oil is shown below:

Lipid profile of some vegetable oils (%)

Fat acid	Sunflower	Corn	Soybean	Cotton	Olive
C.14:0	0,1	0,6	0,2	1,5	-
C:16:0	5,5	13,4	9,5	22,0	11,5
C:18:0	6,0	2,2	3,8	5,0	2,2
T. Sat.	11,6	16,2	13,5	28,5	13,7
C16:1	0,1	0,3	0,2	1,5	0,9
C18:1	31,5	28,6	23,9	16,0	68,8
T. MI.	31,6	28,9	24,1	17,5	69,7
C18:2	49,7	47,7	49,7	50,0	10,5
C18:3	7,3	1,5	7,1	-	0,7

| T.PI. | 57,0 | 49,2 | 56,8 | 50,0 | 11,2 |

Fuentes:

Girasol: Moreiras y cols. (1992)
Maíz: Moreiras y cols. (1992)
Soja: Moreiras y cols. (1992)
Algodón: : Belitz y Grosch (1997).
Aceite de oliva: Moreiras y cols. (1992)

An analysis of this table shows that sunflower, maize and soy oils contain very low levels of saturated fatty acids, similar to that of olive oil, with the exception of cotton, which is made up of these acids. On the other hand, none of them comes close to the oleic acid levels of olive oil, generally less than half. However, as far as polyunsaturated fatty acids are concerned, mainly linoleic acid **(C18:0)** reaches much higher levels, about half of its content and about 5 times higher than olive oil.

It has been proven that polyunsaturated fatty acids such as linoleic acid, show a beneficial effect on the risk factors of cardiovascular disease, and can even act more quickly than oleic acid, however, the stability of these oils is much less than that of olive oil, especially when subjected to heat, in which their oxidation kinetics is much greater than that of oleic acid, and as the main use of vegetable oils is for cooking, and this is associated with the rise in temperature, generally above 100 °C and often much higher than 180 °C in frying, it can be inferred that its benefits are attenuated by the formation of free radicals, peroxides, low molecular mass compounds and high polarity such as aldehydes, ketones and other multiple products that are formed when heated by the reactive lability of the two or more double bonds compared to one in the molecule of oleic acid.

Another negative aspect of the reference oils, in comparison with olive oil, is related to their own nature, as they are products subjected to technological processes of refining, not virgin as they are extracted from plants, so their composition, aroma, taste and presentation are not comparable to that of virgin olive oil. In addition, when they are subjected to refining processes, which involves raising temperatures, and different treatments with chemical products, they alter their initial composition and lose their main bioactive components. On the other hand, as they are treated with organic solvents to increase their yield, there may remain some remnants of these contents in the final product.

At the time we are referring to (mid-twentieth century), olive oil also had another opponent who was forced to leave the lid due to its high content of erucic acid (C22:1:9) that made it unfit for consumption, since this fatty acid hydrocarbon chain greater, is considered a natural toxin. We are referring to rapeseed oil, which, apart from the above, played a dramatic role when lots of oil adulterated with products derived from aniline were sold in Spain, which caused hundreds of losses in human lives and thousands of people still intoxicated with sequels of that unfortunate event.

But locating us in its lipid composition, the original colza contained high concentrations of erucic acid, of the order of 45-54 % of its content, reason why its production was prohibited in the middle of the decade of 1960.

The rapeseed (*Brassica napus*), was practically unknown in the West, although the plant was known from the antiquity in the Asian continent. Its cultivation in Europe is relatively recent, since after the decade of the thirties of last century, motivated by the need for edible oils from plants of cold or temperate climate, during the war conflicts that lashed the old continent, so its introduction into the field of vegetable oils was, and has been, with an overwhelming step, despite incidents related to its composition in fatty acids.

The concentration levels of oleic acid in oilseed rape were between 10-20 %, and it also contained between 5-9 % polyunsaturated fatty acids, and around 6 % saturated fatty acids, but on the basis of its high erucic acid content, close to 50 %, which was considered by international authorities as a natural toxin, the use of rapeseed oil for food purposes of the original varieties was banned in 1965 by the **FDA** (Food and Drug Administration), due to its toxic character, given the high amounts of erucic acid (**C:22:1:9**) it contained.

It seemed then that olive oil lost a competitor, that if it was not relevant at world level, it was in Europe because of the capacity of rapeseed to be harvested in cold and temperate climates, such as those of central Europe, the north of the United States, and Canada, especially in the former, because of its proximity to olive groves. But precisely because of the feasibility of its cultivation in countries with temperate or cold climates, important research was carried out after 1965 to obtain rapeseed varieties with tolerable amounts of erucic acid. In 1974, Canada finally achieved the proposed objectives, so that they achieved a variety of rapeseed that met the expected requirements: climate resistance, high oil content with unsaturated fatty acids, mainly oleic, and above all: low levels of erucic acid. This variety was initially called **L.E.A.R** (Low Erucic Acid Rapeseed), a name that was later changed to canola, which responds to *Canadian oil low acid*.

Canola as a genetically modified species of rapeseed was accompanied by drastic changes in its lipid composition, and if the original rapeseed had very modest concentrations of oleic acid, less than 20%, the new variety raised the same to values equal to or greater than 50% of this acid, while erucic acid levels decreased significantly to figures between 2-5%, ie, the new variety had raised the levels of oleic acid from lower levels of erucic acid, so the bad became good.

These changes led to a significant increase in canola cultivation areas in the cold and temperate regions of the planet until it became possible for its oil to occupy the third place in the world in terms of production and sales, although despite this, olive oil took advantage of rapeseed in two basic aspects: The first is related to the fact that the latter still maintained a certain concentration of erucic acid, and the second to the fact that in the process of obtaining rapeseed oil it is necessary to carry out relatively intense heating, with the possible formation of trans-fatty acids, with which it always remained a suspect in the, so to speak, "*crime scene*". Some of the world's rapeseed fields, around 10%, also maintain the original variety of rapeseed, without undergoing genetic transformation, although located in well-defined places such as India from which it originates and some nearby Asian countries.

Nowadays rapeseed oil is consumed with certain amplitude in the world and its production is several times greater than that of olive oil, slightly higher than that of sunflower oil and below soy and African palm, which occupies the first place in the world.

Anyway, these and other mishaps of the rapeseed or canola, which is the name of the variety obtained by genetic crosses, made that although it could become a certain competitor in the European markets and the north of the American continent, not so in Spain, where although this plant is cultivated, the oil is exported to other countries such as France, and its presence in the markets is made difficult by the syndrome associated with its name with the marked facts of adulteration that accompanied it in the 80's of last century.

From the productive point of view, in addition to the basic character of virgin olive oil as a "flagship" among vegetable oils, it could continue to boast higher levels of monounsaturated fatty acids than other oils, above 70% and more, as well as being accompanied by vitamins, antioxidants and a whole group of substances beneficial to health, although the refined oil of this did not possess them, unless it was mixed, as is generally customary, with virgin oils, all of which gave a respite to this millenarian industry, all of which gave this millenarian industry a respite.

But competition in the oilseeds sector is very strong, almost a war, and refined olive oil, also a product with prices higher than those of its sunflower, maize and soy congeners, gave rise to a competitor from where it was least expected, and not from the basic nature of these oils, but from systematic genetic improvements to reduce the concentrations of polyunsaturated fatty acids in their seeds by increasing, at the expense of these, the levels of monounsaturated fatty acids in what came to be called high oleic oils, in which sunflower and maize stood out more than others, so that oils of this type are already produced and marketed with concentrations of oleic acid similar to that of olive oil. As shown below for high oleic sunflower oil (**HOSO**)

Fatty acids (%)

Oil	Palmític	Steárico	Oleic	Linoleico
Olive	11,5	2,2	68,8	11,5
Sunflower Standard	7,4	5,8	37,3	48,3
Sunflower High Oleic	3,1	5,2	82,2	7,3

Not all High Oleic Oils (**HOOs**) have the same concentration of oleic acid, so there is a wide spectrum of oils of this type, often referred to commercially as frying oils. The name "*high oleics*" is related to the increase in the concentration of this acid in them, which means that they present better indicators for frying such as: boiling temperature, smoke point, thermal stability, and resistance to hydrolysis and oxidation, among others.

The production costs of the **HOOs** are much lower than those of the olive oil, containing the components in monounsaturated fatty acids beneficial for the health that this one provides, nevertheless, is observed with care the properties and the effect of the same ones for the health, given that there is certain controversy on the transgenic character, of hybridization or not, of the seeds of sunflower genetically modified, especially in the countries that regulate the use of transgenic foods as it is the case of those that integrate the European Union.

High oleic sunflower oils show properties similar to olive oils in frying, and are much more resistant to heat and more stable to auto-oxidation processes than the standard oils of this plant, so it is sometimes issued with labels recommending its use for frying, although anything that subjects vegetable oils with high proportions of unsaturated fatty acids to heating processes above certain temperature ranges, or for a prolonged period of time, is poorly recommended for health, given the variety of products that are formed during this process, some of them not recommended to be ingested by cell damage that can cause, given their oxidative ability.

It is necessary to emphasize that when speaking of lipid profile of oils, refers to average values because its composition depends on many factors, including climate, so, for example, in the sunflower crops of certain regions of Argentina standard sunflower oils are obtained with higher concentrations, in 3 or 4 percentage units, than its similar Ukrainian and Russian, for example.

In this competition, the Mediterranean olive oils felt some respite from an unexpected adulteration that occurred in April 2008, in which several European countries, including Spain, detected batches of sunflower oils from Ukraine containing mineral oils as adulterant matter, which led to the governments of these countries demanding the removal of adulterated material and the temporary cessation of the entry of oils from this country.

Subsequently, the European Union was forced to pronounce firmly on this delicate matter, and asked Ukraine for information on the matter and immediate action in relation to this case of adulteration, so this country was forced to solve the problem and ensure the purity of the oils exported, as they are an important commercial item and acquisition of foreign exchange for this country.

In any case, important batches of sunflower oil from Ukraine were withdrawn from the market, although the Spanish authorities considered that the level of contamination did not affect the health of consumers, as pronounced by the then Ministry of Industry in this regard.

High oleic palm oil.

At the end of the last century XX some plantations of African palm (*Elaeis guineensis*) of several South American tropical countries as Colombia, Ecuador and bordering regions of Brazil, saw their cultures infected by the disease known as bud rot which motivated that crosses of this with the American palm *(Elaeis oleifera)*, more resistant to this pathology were made, which led to a hybrid, "**O x G**" (*oleifera x guineensis*), which was not only resistant to the disease in question, but its oil also considerably increased the concentration of oleic acid, and in general that of monounsaturated fatty acids, while decreasing the levels of saturated fatty acids and palmitic, specifically.

With this hybrid, fats with levels of oleic acid equal to and greater than 50% were obtained, while those of palmitic acid decreased significantly and were over 30%. In summary, the following was achieved with this hybrid, without a real affectation of the crop:

-Higher content of oleic acid and monounsaturated fats.
-Lower content of palmitic acid and saturated fats.
-Resistance of the plant to various diseases such as bud rot.
-Lower rate of growth, which implies a lengthening of its useful life of culture.

The high oleic oil obtained from the mesocarp of the fruit of the hybrid O x G, reaches average levels of oleic acid of 54 % compared to 40 % of the African palm oil, as well as a decrease of the content of palmitic acid of 15,5 %, to stand at 28,5 %. The contents of the other fatty acids remained unchanged.

In these oils also reported a higher level of carotenoids in relation to oil palm, while maintaining high levels of sterols and phenolic antioxidants.

The lipid profile of the hybrid palm oil obtained compared to the standard palm oil is shown below:

Fatty acids	Composition (%) of palm oil.	
	Standard	High Oleic
C12:0 Láuric	0,1	0,3
C14:0 Myrístic	1,0	0,5
C16:0 Palmític	43,5	29,0
C18:0 Steáric	4,3	3,0
C18:1 Oleic	**36,6**	**54,0**
C18:2 Linoleic	9,1	12,00

Despite the substantial difference between the two oils, it should not be underestimated that the production volumes of palm oil worldwide are twenty times those of olive oil, and results like these are not at all flattering for olive growers, although for now nothing to fear, as the profile of use of both oils is quite different.

Avocado oil.

Following the doctrines of the old Chinese master of war Sun Tsu on the need to approach the enemy silently to surprise him and try to get the victory, the avocado oil (*American persea*) that was had as an oil more, come to less, or that just began to be mentioned in the sector of edible vegetable oils, has taken a giant leap forward and is being presented as a future rival for olive oil, given its own lipid concentrations, without the need for crosses, as well as other properties that make it have a certain resemblance to olive oil, and that reach such a point, which is an element taken into account by those who are dedicated to the bad art of adulterating oils for profit.

Avocado oil is obtained from the fresh and dried pulp of this fruit, where it is found in varying concentrations, between 5-30 %. It is presented as a golden yellow or amber liquid, soluble in organic solvents and of low viscosity. It has a low saponification index and contains the virgin, significant amounts of sterols and vitamins A, D and E with the consequent antioxidant effect. It is moisturizing and improves skin hydration.

The similarities between avocado oil and olive oil are substantial, even more so in their lipid composition, as shown below.

Average percentage composition of fatty acids of avocado and virgin olive oils.

Acid	Avocado Oil	A. Olive Oil
C14:0	0,03	-
C16:0	16,6	11,5
C16:1	7,1	0,9
C18:0	0,5	2,2
C18:1	63,9	68,8
C18:2	10,8	10,5

As can be seen, without making transformations or drastic genetic changes, there are productive varieties of avocado that produce oils with concentrations that are close to those of olive oil, so it should behave as a protective oil or beneficial for cardiovascular diseases. Although avocado oil contains a slight proportion, less than 6 percentage units, of oleic acid than olive oil, this is counterbalanced by the significant amount of palmitoleic acid it contains, also monounsaturated, and possibly beneficial to the human organism.

As for the main physicochemical properties, a similar correspondence is also established between the two oils, as shown below:

Main physicochemical properties of the virgin avocado and olive oils.

Magnitude	Avocado Oil	Olive Oil
Density	0,910-0,920	0,910-0,916
Ind. Refra.	1,468-1,475	1,4677-1,4705
Index of I2*	85-90	75-94
Index of Sapon.**	177-19	184-196
Index of Perox.***	2-10	20

* cgI2/g
** mg KOH/g
*** meq active oxygen/kg oil

In addition, avocado production has increased significantly in the last ten years, from 3,174 MT in 2003 to 3,444 MT in 2008 and 4,71 MT in 2013. The main producers of this fruit are: Mexico (about 30% of world volume) and other countries such as the United States, Indonesia, Colombia and Brazil, among others.

According to the current production levels of avocado, if this were derived mainly to the production of virgin oil, could be counted on the market with an offer of about 400 000 MT., which would already reach levels above 10% of those of olive oil, accompanied by a growing trend of more than 3%, so that sooner or later could reach it, if current expectations in terms of demand are maintained. However, for the peace of mind of olive growers, the fruit of the avocado, unlike the olive tree, has a wide demand as such for food consumption, given the high qualities of this for direct consumption. Also, and no less important, is that currently the demand for avocado oil is very high in the cosmetics sector. However, we must not lose sight of this competitor which, in a modest and silent way, is making its way into the vegetable oil sector and which shows properties similar to olive oil.

As a conclusion, it is necessary to point out that in the world of oleaginous plants, olive oil has enough enemies to worry about, more than others that may appear, but beforehand it must be pointed out that it is still the most outstanding for its composition, presence, taste, flavour, stability and, above all, for its beneficial character for health, although new products tend to share some of these properties such as high oleic and avocado oils, among others.

Secondary Fraction
of Olive Oil

CHAPTER VI

Secondary Fraction of Olive Oil

Olives.

The fruit of the olive trees: the olive has a high amount of vital nutrients for the human body, which is one of the basic aspects that gives value to olive oil, in addition to the high proportion of unsaturated fats, mainly monounsaturated, omega 9, as the oleic acid it contains. Many of these components remain in virgin or unrefined olive oil, also providing stability and durability to the product.

In general, a study on the average composition of nutrients in the fruit of the olive tree comes close to what is shown below, although these are average values, which may be higher or lower depending on the variety of olive and other conditions associated with the crop.

Components of the fruit of the olive tree*

In 100g of fruit there is on average:

Fat: 16,3g

  -Saturated: 2,03g

  -Monounsaturated: 11,3 g

  -Polyunsaturated: 3,03g

Carbohydrates: 4,40 g

-Sugars: 0,55 g

-Fiber: 3.85 g

Protein 1,03 g

Vitamins:

A (Retinol): 20 µg

β- carotene: 231mg

Thiamine (B1): 0,021 mg

Riboflavin: (B2): 0,007 mg

Niacin (B3): 0,237 mg

B6: 0,031 mg

Folic acid: (B9): 5 µg

E (tocopherols): 3,81mg

K: 1,4 µg

Minerals:

Na: 1,56g

K: 42 mg

Fe: 49 mg

P: 4 mg

Ca: 52 mg

Mg: 11 mg

Water: 50

*Based on data from the USDA (United States Department of Agriculture).

In accordance with the above, and since virgin olive oil is obtained by direct extraction of the fruit of the olive tree without being subjected to any subsequent process, only to the corresponding physical separations, it is to be expected that most of these beneficial components for health are incorporated into virgin olive oil to give it that differentiating touch in the nutritional aspect, which makes it the most complete of edible vegetable oils as indeed occurs.

However, this does not happen with all these components, because when separating the water in the obtaining of the oil, the soluble matters in it are separated and they do not remain in the oil, or their quantity is much smaller, reason why it will be treated next on the properties and characteristics of the substances that in the end are maintained in the product in quantities, that if not appreciable, can exert influence on the properties of the same one, its stability, its resistance to the oxidation and its nutritional qualities.

In this chapter we will evaluate the secondary components that accompany virgin olive oil in what is called the non-saponifiable fraction, a little less than 2 %, which because it comes directly from a fruit has numerous components beneficial to health such as, among others: squalene, β-carotene, chlorophyll, tocopherols, sterols, and polyphenolic compounds; the latter with a marked antioxidant action, evidenced by the high stability of virgin olive oil once packaged.

Some components of the secondary fraction of olive oil are listed below.

Components Concentration mg/kg oil*

Terpenic alcohols 3500
Sterols 2500
Hydrocarbons 2000
Squalene 1500
Phenolic compounds 350
Beta carotene 300
Aliphatic alcohols 300
Tocopherols 150
Esters 100
Aldehydes and ketones 40

---

* Mataix J. (2001) Virgin Olive Oil: Our food heritage. Granada University.

Seen in this way, virgin olive oil behaves in an intermediate way, as a vegetable oil to which the properties of the olives are added, so that the varieties of the latter affect the characteristics of the oil obtained, as well as the form of harvesting and processing, and of course the conditions and time of storage.

Virgin olive oil is shown to be relatively more acidic than refined seed oils, which responds to the presence of free fatty acids typical of olives, or formed during the harvesting and manufacturing processes, which has nothing to do with taste and other organoleptic aspects, although they are an element to be taken into account in their classification as extra virgin, virgin or glaring.

The unsaponifiable components of olives are not eliminated in the mechanical process of obtaining virgin olive oil, but to a large extent in that of refined oil, so that a clear dividing line is established between these types of oils, as if they did not share their own nature. In this sense, except for its lipid composition, refined olive oil is more similar to seed oil, especially to high oleic and canola oil, enriched in oleic acid by transformation and/or genetic selection of the seeds.

Unlike other oils, the compounds that accompany virgin olive oil, responsible for its organoleptic properties, play a crucial role in determining its quality and classification.

Virgin olive oil includes a wide variety of vitamins, including vitamins A, E, and K, as well as steroids such as sitosterol. It is considered to be one of the highest concentration of tocopherols (vitamin **E**) contained within edible oils, which gives it a high antioxidant power, maintaining its stability and composition over relatively long periods of time, so that it can be consumed without showing appreciable alterations up to a year after packaging and without the need to add natural or artificial antioxidants.

It has been reported that the virgin olive oil contains numerous group of substances of varied composition, foreign to those corresponding to its lipid profile. These, although secondary and in a proportion of around 2% play an important role in this oil and define in many cases its nature and beneficial action for the body. Among them they stand out: *

Terpenes:
Squalene: 300-700 mg / 100 g
Carotenes: 0.5-10 mg /Kg

Chlorophyll: 0 to 9.7 ppm

Tocopherols: 7-30 mg / 100 g, prevailing the
γ-tocopherol with an average concentration of 93 %.

Sterols: 80-240 mg/100g
Campesterol: 2.0-3.0 %.
Stigmasterol: 1.0-2.0 %.
Avenasterol: 95.0-97.0 %.

Phenolic compounds: 50-500 mg/Kg

Other organic compounds to a lesser extent:
Alcohols
Ketones
Aromatic esters
Ethers
Furanic derivatives, among others

*(Source: Mataix and Martínez de Victoria (1988).

All these components that show a remarkable biological activity are beneficial for the human organism, characterize olive oil and make it stand out above other vegetable oils, as well as its lipid composition rich in oleic acid. It is true that virgin oils extracted from other oleaginous plants also have substances in their composition, in some cases, similar to those of olive oil, but most are not suitable for direct consumption, or their organoleptic properties do not make them appetizing to be ingested in the form in which they are presented.

Of the secondary components of olive oil, it is necessary to highlight the high antioxidant power of some of them, such as tocopherols, polyphenols and vitamin A, which in addition to translating into practice in raising the stability of the product, avoiding or delaying oxidation and rancidity, similarly exert a cell protective role in the human body, delaying or inhibiting the biological oxidation of cells avoiding the disorders related to it.

Squalene ($C_{30}H_{50}$).

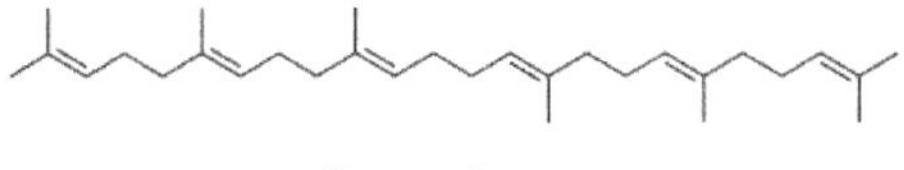

Squalene

Pale yellow translucent liquid

Melting temperature, 0,5 °C

Bt.275 °C
Density 0,858 g/cm³
M: 410.72 g/mol

Squalene is the main component of the unsaponifiable fraction of virgin olive oil and is present in virgin olive oil in concentrations between 60-75 %. Its structure corresponds to that of an aliphatic hydrocarbon with conjugated double bonds. It is a precursor of the biosynthesis of sterols among which are cholesterol and vitamin D, among others. It contains 6 units of isoprene, so it is possible that these are formed by cicloadition in the bonds 1.4.

Squalene belongs to the group of carotenoids, within which is the β-carotene, which is also found in carrots and other fruits and vegetables causing the colour of them. The β-carotene is oxidized in the liver to produce vitamin A ($C_{20}H_{30}O$), so it is considered a precursor of this known action in the formation of images in the retina, in addition to many other beneficial properties for the body.

Clorophyl.

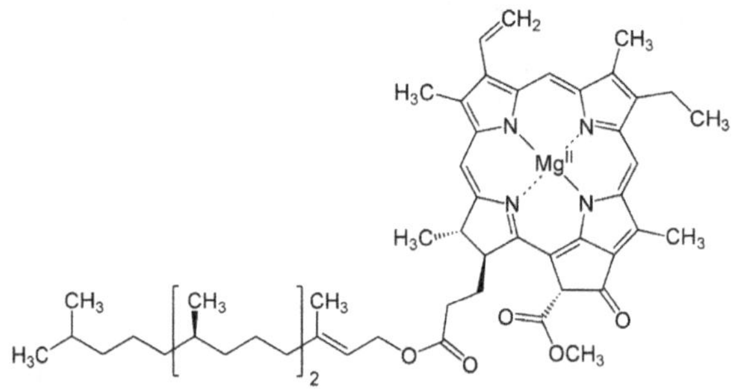

Structure of chlorophyll a

Chlorophyll is a vegetable pigment of singular importance for plants because it is the one that conditions the photosynthetic processes that are carried out in it, and therefore the production of sugars and starches from water and carbon dioxide through sunlight.

Before reaching maturity, many fruits such as olives show a green hue due to this pigment, which passes to the oils when they are extracted from the fruit. As this pigment decreases its concentration as the fruit ripens, the oils may suffer a colour variation ranging from dark green to tenuous yellow. For this reason, when the harvest is early, the virgin oil presents a higher concentration of chlorophyll and takes on a dark green colour, which becomes lighter and dimmer with the progress of the harvest.

The chlorophyll present in virgin olive oils degrades over time through a process accelerated by temperature and exposure to light, so the oils are changing the shade of green for yellow, although this is not an organoleptic indicator during the tasting, and generally professionals dedicated to this work, use blue glasses to not be affected subjectively their criteria.

The green colour of chlorophyll responds to its maximum absorption in the wavelength range of 400-500 nm (blue) and 600-600 nm (red), the combination of which corresponds to a spectral mean centre 500-600 nm, i.e. the green colour that identifies the human eye. The intensity of these bands is correlated by Lambert-Beer's law with the concentration of chlorophyll. When chlorophylls are transformed over time into carotene, the bands associated with the yellow colour appear.

Structurally, chlorophyll consists of two parts: a porphyrin ring with coordinated links to Mg whose function is related to the absorption of sunlight and a phytol chain necessary to keep the pigment attached to the photosynthetic membrane.

There are different forms of chlorophyll that it is not necessary to value in this work, but it is necessary to emphasize that to her are attributed different beneficial properties for the health as her antioxidant, nutritional action, and also there are references of her lipid-lowering and antimutagenic action, among other qualities, reason why her presence in the virgin olive oil gives a supplementary value to this oil.

Vitamin A: Retinol. ß-carotene.

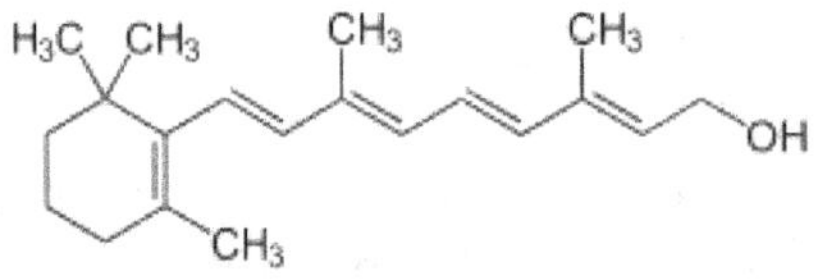

Retinol

It is a fat-soluble vitamin that appears in vegetables in the form of carotenes and in animals in their exact and natural form. Its protective and beneficial effect for health is given by the facility to capture radicals and free oxygen. It is present in many varieties of plants to which it communicates its colour. It is very unstable to heat and metals from cooking utensils such as iron and copper, among others.

Carotenes are considered to have antioxidant action because of their ability to sequester free radicals, including atomic oxygen.

Tocopherols.

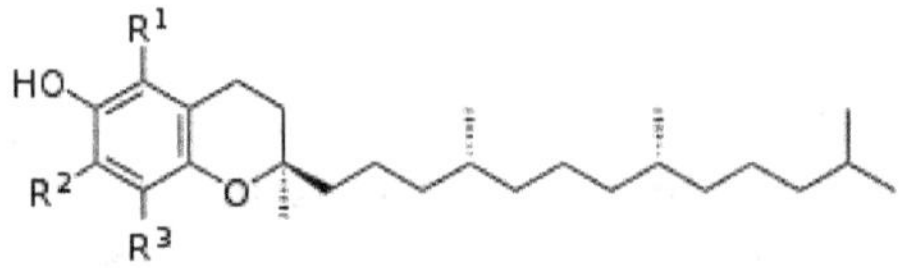

General structure of tocopherols

Tocopherols are one of the basic components of olive oil and give it most of its antioxidant properties. Their quantities vary depending on the variety of olive from which the virgin olive oil is extracted, as a substantial part of it is lost during the refining during the technological process to which it is subjected, which makes it necessary to add the lost or necessary quantities at the end in order to maintain its stability. This, as has been explained, does not occur in virgin olive oil which contains only those extracted from the fruit during milling, but which are sufficient to prevent oxidation and deterioration over a long period of time.

In olive oil, alpha, beta and gamma tocopherols have been identified to a greater or lesser extent, although γ-tocopherol generally prevails. These substances are very important for the organism because thanks to the OH group present in the aromatic ring they can capture free radicals and reduce the kinetics of cellular oxidation. The alpha isomer (**α**) known as vitamin E and plays an important role in the organism. The properties and characteristics of some of them will be studied below.

Vitamin E (α-tocopherol): Tocopherols and tocotrienols.

They are lipid antioxidants of great efficiency to capture oxygen, limiting the formation of peroxides in the cellular metabolism of lipids associated with unsaturated fatty acid molecules.

Chemically, these compounds are polyprenoids characterized by the presence of an aromatic ring with a hydrophilic group and a polyprenoid chain. If the chain is saturated it corresponds to tocopherols, if it is unsaturated it corresponds to tocotrienols.

α-tocopherol

α-tocopherol is presented as a white solid with a molecular mass of 430.7 g/mol and a density of 0.95 g/cm³, not very soluble in water, but it is soluble in oils and other low polarity liquids.

In its mechanism of action, α-tocopherol acts by preventing the oxidation of fatty acids and therefore the formation of peroxides. Of the tocopherols it is the most active as an antioxidant agent.

Gamma tocopherol

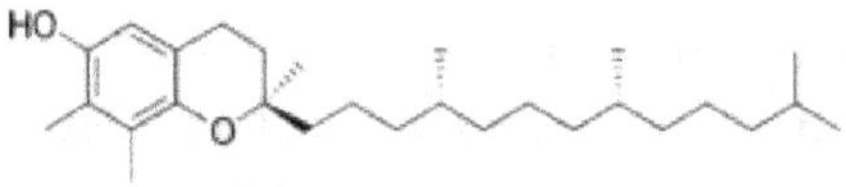

γ-tocopherol.

It's one of the forms of vitamin E. It is commonly presented as a slightly pale yellow, oily liquid. Its molecular mass is 416.7 g/mol and its solubility in water and polar liquids is very low, but it dissolves well in less polar organic solvents such as ethanol, acetone and vegetable oils.

Since it is relatively not soluble in water, but it does so in low polarity liquids such as lipids, it is very useful for slowing down the oxidative degradation of fats and thus preventing them from becoming rancid. It is an effective antioxidant present in olive oil and other fats of vegetable and animal origin.

δ-tocopherol.

It is presented as an oily liquid with a certain viscosity, of a tenuous yellow tone, with a molecular mass (M) of 402.7 g/mol, not very soluble in water, but slightly soluble in less polar liquids such as vegetable oils. It is one of the forms in which vitamin E is present.

It exerts a strong antioxidant action in olive oil, although its antioxidant activity is slightly lower than its isomers.

Vitamin K. Phylloquinone.

Vitamin K (C$_{31}$H$_{46}$O$_2$)

2-Methyl-3-[(2E)-3,7,11,15-tetramethylhexadec-2-en-1-yl]naphthoquinone

M: 450,70

Vitamin K, also known as phylloquinone, is a substance with coagulating or anti-haemorrhagic properties widely used in the health sector to prevent cases in which there is a risk of bleeding during surgery of greater or lesser magnitude. It is a fat-soluble substance found in both the fruit and the oil of olive trees. According to its structure, it is a compound derived from 2-methylnaftoquinone.

All forms of vitamin K share a methylated ring of naphthoquinone in their molecular structure, which may vary according to their aliphatic substitutes in position 3 of the chain. Phylloquinone, a natural variant of vitamin K, contains four isoprenoid residues in its lateral chain, one of which is unsaturated.

Vitamin K is found in olive oil to a lesser extent than vitamin E, but plays an important role within the properties of this oil. It is soluble in lipids, but not in water, so it remains in the oil once the juice has been extracted from the olives and moisture has been separated.

It is also known the action of vitamin K in the generation of red blood cells.

Terpenic compounds.

The concentration of terpenic compounds is one of the highest within the unsaponifiable fractions of olive oil and may occur showing an alcoholic function with tetracyclic (uvaol) or pentacyclic erythrodiol structure, including acid such as the corresponding oleanolic and maslinic acids. As uvaol and erythrodiol are found in greater abundance in the skin of the fruit, they concur in greater presence in the olive marc oil and their measurement can be an adequate indicator to detect adulterations of virgin olive oil by means of olive marc oil, so the maximum possible concentrations of these compounds in virgin oil are regulated.

Erythrodiol ($C_{30}H_{50}O_2$)

Uvaol ($C_{30}H_{50}O_2$)

Other terpenic compounds found in virgin olive oil in a less representative way are some triterpenic alcohols with a pentacyclic structure such as cycloarthenol and amirines α and β among others.

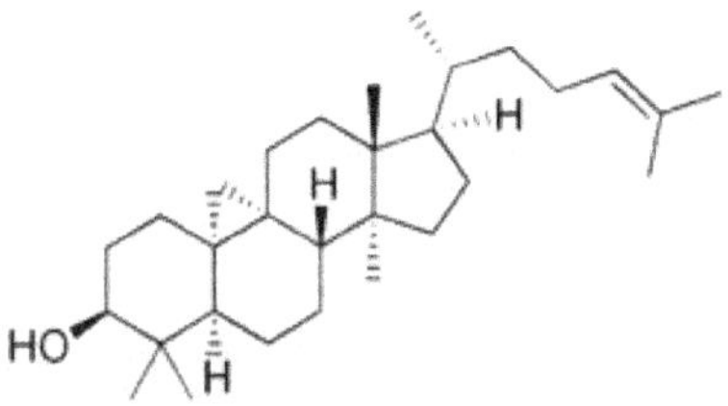

Cycloarthenol ($C_{30}H_{50}O$)

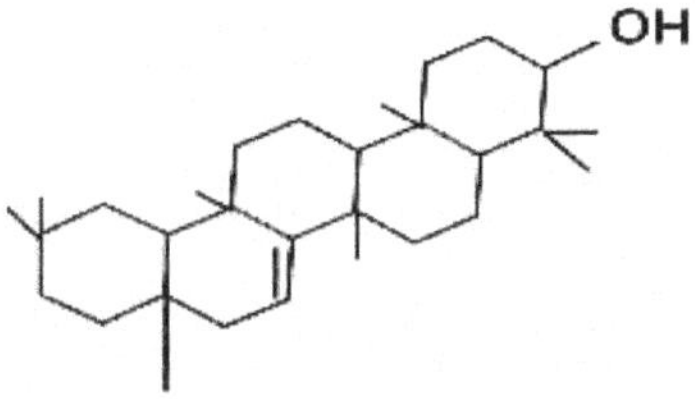

β-Amirine ($C_{30}H_{50}O$)

Phytosterols.

These compounds are presented in olive oil under the preferential form of sitosterol and avenasterol, they come to be like the homologues of the cholesterol in the plants. The most abundant in virgin olive oil is sitosterol in three or four times the amount of avenaesterol, in smaller proportion is also campesterol and stigmasterol, the latter can be an element indicative of adulteration of virgin olive oil when it is in quantities above normal, since in other oils such as sunflower oil contains it in much more significant quantities. The content of total sterols is higher in virgin olive oils than in refined oils, as this is water-soluble and is lost during the technological process of refining.

β-Sitosterol C₂₉H₅₀O)

Δ5-Avenasterol C₂₉H₅₀O)

Campesterol ($C_{28}H_{48}O$)

Solid state at room temperature.

M: 400.68 g/mol

T.fus. 157,5 °C

It is a stereoid derivative with a molecular structure similar to that of cholesterol, and at the same time the simplest sterol, has an OH group in position 3 of the stereoid structure or skeleton, with sigma or saturated bonds in the rest of the molecule, except for a double bond in the second ring. It has cholesterol-lowering activity by limiting or competing with cholesterol in its absorption by the small intestine. Although it is found in lesser amounts in olive oil, other vegetable oils also contain it, such as soybean oil, for example. Together with stigmasterol and brassicasterol, a hypocholesterolemic complex drug is produced. For this reason phytosterols are used as food additives in some fatty products such as margarines and butter. It is also conferred anti-inflammatory action.

Long chain aliphatic alcohols

Olive oil also contains aliphatic alcohols with an even number of carbon atoms in hydrocarbon chains of 18 to 28 atoms, the most abundant of which are hexacosanol and octacosanol, although several others can be found in insignificant quantities. These, as well as sterols and terpenes can be found in esterified form with different fatty acids.

Hexacosanol. Hexcosan-1-ol

$C_{26}H_{54}O$

$CH_3(CH_2)_{25}OH$

It is presented as a waxy solid at room temperature and shows a chain length of 26 carbon atoms. It is soluble in organic solvents but not in water. It is more abundant in the cover of the fruit.

M: 382,7 g/mol

Melting temperature: 79 ° C

Bt. 240 °C

Long chain esters.

In addition to the triacylglycerides formed by fatty acids with glycerin, the other alcohols present in olive oil can form esters with fatty acids that are in a free state, which opens up a wide range of possibilities and makes this oil as rich in different and varied compounds as the esters referred to above. Among these, we can mention the waxes obtained by joining the alcohols of high molecular mass (C20-C28) with the fatty acids of long chain to form waxes with more than 40 atoms of carbon and of even number, which are the ones that are more found in the olive oil and that pass to this one during the extraction process, since they are in the skin of the fruits to protect them and to avoid their loss of water or dehydration.

Phenolic compounds.

These important compounds with a high antioxidant action differ somewhat from the compounds contained in the secondary fractions of the olive oil studied so far, as they have a higher polarity and their content in the oil depends on different factors, including the degree of maturity of the fruit subjected to extraction in the oil mills, and countless other factors, including climatic ones. In the refining process, a large quantity of these valuable substances is lost, so their presence in virgin olive oil is one of the aspects that gives value and importance to this product. They are found in a variable amount in this, approximately between 50-200 mg per kg of oil, although this can vary substantially to much greater or lesser quantities.

As expressed, polyphenols are found in appreciable quantities in virgin olive oil and are among those that contribute most to the anti-oxidative stability of this, are also largely responsible for its bitter taste. It has been proven that the stability of oils against

96

oxidation is directly proportional to the concentration of polyphenols present. To reach this conclusion, analytical techniques such as *Rancimat* were followed, that is, to measure the time necessary for an oil to begin to present rancidity qualities when it is heated to temperatures equal to or higher than 100 °C.

The polyphenols found in olive oil vary in content depending on the type of olive, the growing conditions, as well as their own characteristics and the way they are stored and preserved.

Polyphenols are slightly acidic and water-soluble, which means that they are found to a greater extent in virgin olive oil than refined olive oil, so they must not be present, or in very small quantities, in refined seed, marc or olive oils. These polyphenols are contained in the fruits of the olive tree and pass into the oil during the milling and extraction process. In general, the concentration of polyphenolic compounds is a measure of the quality of virgin olive oil.

The phenolic compounds found in greater proportion in olive oil are: oleuropein - to which reference will be made later -, tyrosol and hydroxytyrosol, as well as others of lesser complexity.

The stability of the oils and some organoleptic aspects respond to these compounds, they are very antioxidant, the most important being hydroxytyrosol.

Tyrosol (4-(2 hydroxyethyl)phenol).

$C_8H_{10}O_2$

Physical state: solid

M: 168.16 g/mol
Density: 1,2 g/cm³
Melting temperature: 90 °C
Boiling temperature: 287 C° to 760 mmHg
Refractive index: 1.578

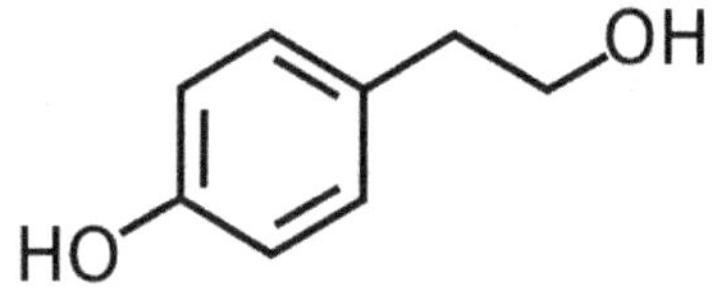

Tyrosol

It is a phenolic derivative found in certain plants and, of course, in olive oil and argan oil. It has antioxidant properties derived from the two OH groups that it present, one of them attached to the ring, which gives it phenolic properties and the other to an ethyl chain. It is less powerful than other phenolic derivatives, but at the same time it is more abundant, which attenuates this effect. It is considered a cellular protector against oxidation.

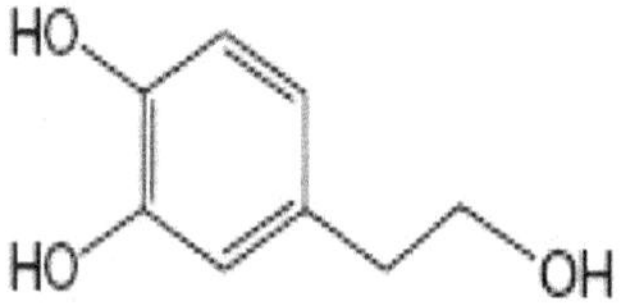

Hidroxitirosol

In addition to this simple phenolic derivative, others can be mentioned such as p-cumeric acid, gallic acid, etc., in a wide variety of products with a greater or lesser antioxidant effect, it is also worth mentioning derivative esters such as tyrosil acetate, among others.

The presence of flavonoids such as apigenin has been recorded with a polyphenolic character in a very wide variety of types and structures, generally derived from simpler molecular forms such as:

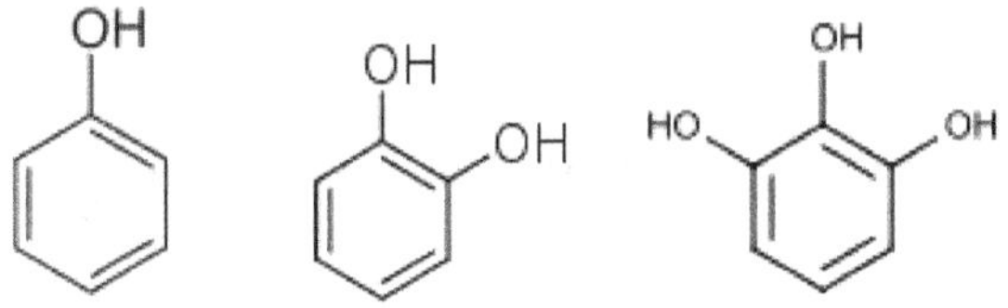

Phenol   Pirocatechol   Pirogalol

Other polyphenolic compounds of media interest.

In recent times different compounds have been discovered or identified in fruits and olive oil, with interesting properties beneficial to health, some even with potential pharmacological effect, within these will highlight some of them as the *oleocanthal* and *oleuropein,,* among others, as well as the results of research carried out so far.

Oleocantal

(Carboxymethyl ligustroside aglycone)

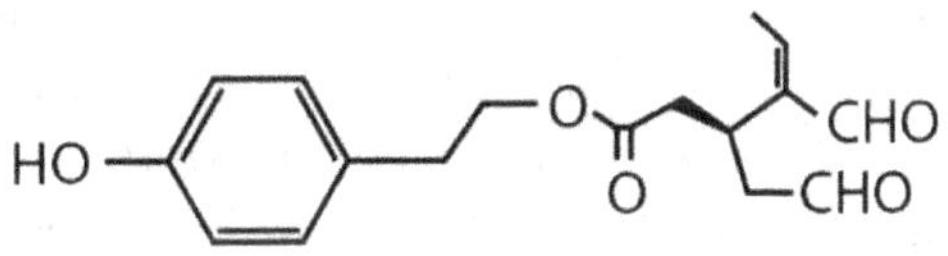

Oleocanthal

It is a natural substance characterized by giving, or intensifying the harsh, bitter and spicy flavour to the extra virgin olive oil, or of superior quality. It was discovered in recent years in some Italian olive oils, but in even more recent studies it is considered to be more frequently found in Greek olive oil batches. It is not present in all olive oils, at least in appreciable quantities.

The interesting thing about this substance is that it shows effects similar to those of non-steroidal anti-inflammatory drugs such as ibuprofen, without being accompanied by the contraindications of this, given that it is integrated in virgin olive oil in small proportions.

The term oleocanthal is related to the properties of the compound:

Oil: oil, canth: spine, and the final suffix "al" corresponding to the aldehyde function.

Chemically speaking, the *oleocanthal or oleocantal*, as it is often written in Spanish, is a tyrosol ester with a structure derived from oleueropein, a compound also identified in olive oil. Its structure is, of course, different from that of ibuprofen.

In its first trials, its irritant action on the throat was measured and positive correlations were found between the degree of throat and the concentration of the product.

The pharmacological actions of oleocanthal are related to its anti-inflammatory and antioxidant effect, by interfering in the metabolism of prostaglandins, limiting, or attenuating the action of the enzyme cyclooxygenase, without doing so selectively. Other recent studies indicate that it also exerts an antiproliferative effect on tumour cells, favours their cell death through caspase-3 and also shows anti-migratory and anti-invasive activities. It is also associated with positive effects on neurocerebral diseases related to Alzheimer's disease (studies in "vitro" and in mice), but given the short time that has elapsed since the discovery of this substance, barely 15 years, this prevents definitive conclusions from being drawn in this respect.

The average concentration of the oleocanthal in the olive oils studied is of the order of 50 mg/L, so an intake of 30 ml of oil per day, with an absorption of between 60-90 %, could provide the individual with a daily dose of 5.4 mg of this natural drug, useful for the prevention of different diseases, similar to studies that have been carried out with acetylsalicylic acid (aspirin) which also interferes in the synthesis of prostaglandins.

Oleacein

It is a compound of structure similar to the oleocanthal, with the only difference that shows a group OH attached to the ring in ortho position with the original OH of this one, so that it has in total two groups OH in this ring, unlike the oleocantal that has one.

It has been shown that oleacein inhibits the action of 5-lipooxygenase, important in the treatment of respiratory diseases such as asthma and other inflammatory processes. Studies carried out to determine the composition of this substance indicate that it is found in concentrations of 65 ppm in olive oil, half of the content of oleocanthal and the third part of aldehyde polyphenols in the oil.

Oleuropein ($C_{25}H_{32}O_{13}$)

Oleuropein

M: 540,51 g/mol

It is a dryiridoide glycoside sterified with a phenylpropanoid alcohol, which is incorporated into the virgin olive oil from green olives, and gives a bitter taste to this oil. In addition to the pulp, the oleurepein is found in the leaves of the olive trees. This compound has also been identified in argan oil.

It can be inferred that oleuropein is a powerful natural antioxidant, only to realize the large number of OH phenolic groups it presents in its structure, so it helps the stability and durability of olive oil, as well as its resistance to rancidity and oxidation.

The pharmacological action of oleuropein has been studied in experimental animals from which conclusions are drawn that injected intravenously reduces blood pressure and dilates the coronary arteries, on the other hand, in vitro studies suggest that inhibits the oxidation of LDL cholesterol, so it can have a positive effect on CVD.

Oleuropein can be converted into elenolic acid during the saline processing of olives, which preserves them from bacterial attack and favours their preservation. Hydroxytyrosol derived from oleueuropein is a powerful antioxidant agent.

Volatile Compounds.

As in other vegetable oils, in olive oil there are volatile substances of low molecular mass, which can pass to the vapour phase at room temperature, and above all, under heating, causing flavours, smells and aromas in it. Nearly 100 of these substances have been identified, although some are formed during the oil storage process, since the fruit is processed as soon as it reaches the mill. Its nature and composition depends on numerous factors, including the nature of the olives, the state of ripeness, climatic conditions and cultivation, among others.

The nature of these volatile minority compounds in olive oil responds to different organic functions within which they have been characterized: alcohols, aldehydes, acids, esters and phenols among many others. The most abundant and frequent oils obtained in the Mediterranean Basin are hexanal, hexanol, 3-methylbutanol, among others, as well as trans-2-hexenal unsaturated aldehyde.

B-complex Vitamins.

Although their existence in olive oil is very limited given the generally water-soluble nature of these vitamins, they do exist in the fruit of the olive tree in approximate amounts of 3 mg/kg of olives, within which thiamine (**B1**) has been identified: 0.21 mg/kg; riboflavin: (**B2**): 0.07 mg/kg; niacin (**B3**): 2.37 mg/kg, the most abundant, and B6: 0.31 mg/kg.

Thiamine (B1). ($C_{12}H_{17}N_4OS+$).

Thiamine.

Molecular mass: 365 g/mol.
Melting temperature: 248 °C

From the chemical point of view, this molecule is formed by two cyclic structures linked through a pyrimidine ring with an amino group, and a thiazole ring joined to the pyrimidine by a methylene bridge.

This compound is also known as thiamine and is part of the known B vitamin, it is soluble in water and glycerin, but not very soluble in less polar solvents such as ethanol. Therefore, during the production process of virgin olive oil, most of it can go with the water, leaving very little or minimal proportion in the final oil.

Vitamin B1 is well known for its lack in the body causes diseases such as beriberi, and Korsakoff syndrome.

Thiamine plays a fundamental role in the oxidation of carbohydrates, with the release of energy necessary for the functioning of the body. It also has an impact on the nervous system.

Riboflavin: (B2) ($C_{17}H_{20}N_4O_6$).

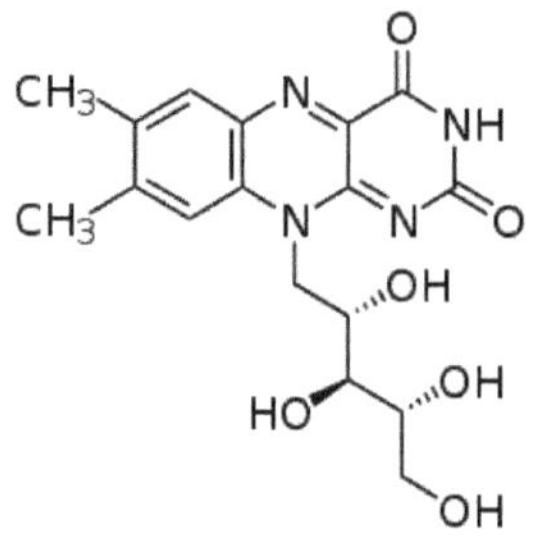

Riboflavin

M: 376.36 g/mol
Melting temperature: 280 °C

It is presented as a yellow solid soluble in water, and consists of a dimethylated isoaloxazine ring attached to the ribitol with its chain of 5 carbon atoms. Like thiamine, it plays an important role in the energy metabolism of carbohydrates and also in that of other biomolecules such as lipids, and proteins. It is sensitive to sunlight and heating. The lack of this in the organism causes ocular disorders, cutaneous and fatigue, among others.

Niacin (B3) ($C_6H_5NO_2$)

Pyridin-carboxylic acid

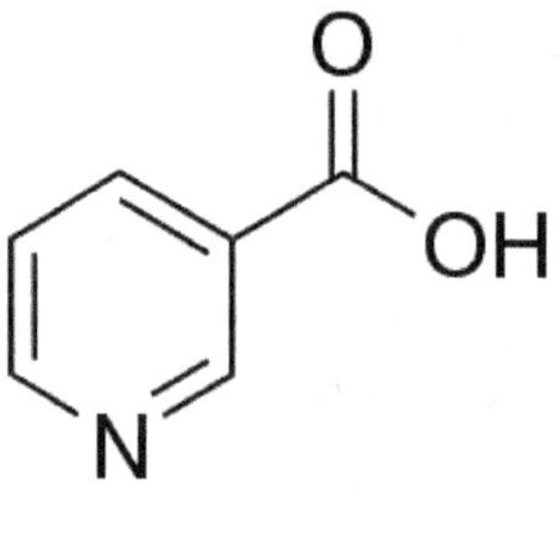

Niacin

M: 123.11 g/mol
Melting temperature: 237 °C
pKa: 4,87

In spite of being found in small quantity in the olive oil it is the most abundant of the vitamins of the complex B found in it. It is soluble in water and has an acidic character, as its name indicates. It is also known as nicotinic acid.

B3 also includes amide derived from nicotinic acid: nicotinamide ($C_6H_6N_2O$).

This substance performs important functions in metabolism such as the elimination of toxic substances harmful to the body and participates in the synthesis of steroid hormones by the adrenal glands. It also acts on the cellular metabolism integrated in the coenzymes **NAD** and **NADP**. It participates in oxidation reactions of carbohydrates, lipids and proteins. It also affects the functioning of the nervous system, as well as the circulatory system.

Vitamin B6

It is actually a group of three chemical substances of very similar structure: pyridoxine, pyridoxol and pyridoxal.

Pyridoxine   pyridoxol   pyridoxal.

The basic difference between the three structures is derived from the substituent groups attached to the pyridine ring: in the pyridoxine alcohol derivatives, in the pyridoxol one of them carbonylic and in the pyridoxal one of the groups is amino.

They are water-soluble compounds and the pyridoxal and pyridoxamine phosphates function as coenzymes in different enzymatic reactions related to the metabolism of amino acids, in which they deal with the transfer of the amino group (transaminases).

Its deficiency is rare in the organism, unless people present nutritional problems related to a deficient diet, as generally occurs in poor countries with low economic development index.

Pyridoxal phosphate serves as a coenzyme in the metabolism of neurotransmitters that regulate the state of mind and in the synthesis of dopamine, adrenaline, etc., as well as γ-aminobutyric acid acts as an inhibitory neurotransmitter very important for the functioning of the brain.

Vitamin B6 is very common in the sports world due to the increase it can cause in muscular performance and energy production, a basic aspect in these activities, given that they favour the release of glycogen by the liver.

Its deficiency in the organism, very rare in the population of developed countries that have a normal diet, is manifested by neurological abnormalities: peripheral neuritis, as well as pain in the extremities. In underdeveloped countries with food limitations, it occurs much more frequently.

Virgin olive oil, additives.

It is increasingly common to find virgin olive oils in supermarkets and retail outlets to which certain types of ingredients have been added to give them flavour, aromas, etc. These types include oils flavoured with lemon, chilies, basil, etc. This, however, does not comply with the regulations established by the International Olive Oil Council (**COI**), concerning the regulations governing olive oils, which determined that at its 22nd Extraordinary Meeting this body made a reminder and made a warning that such actions are not permitted, at least according to the definitions of virgin olive oils, which exclude the addition of any type of additive, and to say any, is any, that is, none.

The addition to virgin olive oil of any of the related ingredients or others of a different nature immediately leads to the oil losing its virgin name and it would therefore be more advisable to call them olive oil with garlic, rosemary, etc., dry, without the use of the term "*virgin*".

In the case of refined olive oils and marc oils, it is possible to use the additive **E307,** α-tocopherol, as an antioxidant in the proportions that this may decrease during the refining process, without exceeding the maximum dose of 200 mg/kg of α-tocopherol total in the final product.

In essence, virgin olive oil is defined and must be so, as *virgin oil*, without any type of manipulation or transformation, as the natural product obtained from the crushing of olives, without additive of any kind (preservative, colouring or chemical agent).

FURTHER READING

Olive Oil and Mediterranean Diet

Today, the Mediterranean diet is one of the most important ways of supplying healthy and healthy food and nutrients to the human organism and is closely related to the eating habits and forms of the inhabitants of the countries of the Mediterranean Basin. Precisely, as a result of the studies carried out by Keys and collaborators in the mid 1950s on the protective role of unsaturated fatty acids on atherogenic risk factors, this diet came into force and also as evidence was found that in these countries was much lower the rate of aging than the world average and were recorded less frequently cases of cardiovascular disease.

Without a doubt, apart from the methods and ways of executing the Mediterranean diet and its composition, virgin olive oil is the basic element of it and without it we could not even talk about it.

Indeed, the nutritional properties of virgin olive oil derived from its lipid profile rich in monounsaturated fatty acids of the omega 9 series, and its secondary components with a high number of compounds with biological activity and protectors of the human body, make this oil an indispensable element of the Mediterranean diet and is closely related to the high indicators of quality of life of the inhabitants of the Mediterranean region.

The hypocholesterolemic action of olive oil as a protective agent against cardiovascular diseases is the basic element that determines the lower incidence of this serious disorder in the inhabitants of the Mediterranean Basin, which has been evidenced in the results of numerous clinical trials in which it has been possible to quantify the favorable changes of subjects in studies in key indicators such as total cholesterol (**Colt**), low-density lipoproteins (**LDLc**) and high-density lipoproteins

(**HDLc**), decreasing the first two which are risk factors for atherogenic damage and raising the third which exerts a protective action on the body against this disease.

Among the components that make up the Mediterranean diet, virgin olive oil is very close to the base of the pyramid, only, without any other accompanying oil or fat, so it has the responsibility within this diet to provide the basic lipid ingredients that the body needs to perform its functions.

The food pyramid that corresponds to the Mediterranean diet is composed of different nutritional elements, including: fruits, vegetables, nuts, carbohydrates in the form of cereals, dairy products, eggs, and fish, the latter as a source of omega 3 fatty acids, and to a lesser extent red meats and of course virgin olive oil. All of them play an important role in nutrition and have a certain position in the food pyramid, but very close to its base, due to its importance, is the virgin olive oil, which provides numerous components necessary for the proper functioning of the body and which in one way, or another, were treated in this last chapter.

According to the data on life expectancy in the world exposed by the World Health Organization in 2018, three countries of the Mediterranean Basin occupy a prominent place in this list: Spain in third place with an average life expectancy of 83.1 years, Italy in 7th place with 82.8, while Greece, which is another major consumer of olive oil, ranks 23rd with a life expectancy 81.2 out of a total of 183 countries. The reader can draw his own conclusions from this.

OLIVE OIL
ADULTERATION

CHAPTER VII

Adulteration of Olive Oil

Not a year goes by without the publication of new cases of adulteration, or attempts to adulterate batches of olive oils, especially the so-called "extra virgin"; and not only in the countries of the Mediterranean Basin, also in the United States of America and in other countries that carry out this production on a smaller scale, so it seems that it is an evil that is practised all over the world. The reason for this, without exonerating the offenders, seems to be given by the high value of this oil on the market, and the difficulties in quality control, by referring to items valued by tasting through their organoleptic properties.

This fact involves the agents involved in the handling of olive oil, including producers and distribution markets, which could apparently be the victims of these bad practices. Proof of this is the study carried out by the **OCU** (Organización de Consumidores) in 2011 in Spain on a sample of forty brands of oils distributed to consumers, in which eleven of them detected anomalies of different types, including that some were not even fit for consumption.

This study, questioned by some large companies of traditional brands involved, for considering that it is based on subjective evidence related to tasting, or that the anomalies could be a consequence of time and storage conditions, does not in any way demerit the study of a neutral body that watches over the safety of the consumer, and that enjoys a well-earned prestige among the Hispanic population.

Apart from the results: subjective or not of the **OCU**, it is found that the same can occur in the same way as the other side of business and mercantile, because the methods and techniques are the same, not determined by analytical laboratory tests, taking into account the characteristics of the product that is sold.

In the matter in question, supermarket chains related to adulterated oils were also involved, or those that did not meet the requirements for sale and that had consciously or unconsciously oversized the price that should be established according to their degree of quality. What is *virgin, glaring* or *refined* cannot be sold as extra virgin, since the price difference between these is very significant and much less in the case of oil qualified as *glaring* not considered suitable for direct human consumption.

Guilt cannot be attributed to time, or storage conditions because this is fully regulated and alterations to these conditions cannot affect the consumer. And if the product had suffered some kind of deterioration in its storage, or had been overdue, this could not be distributed to the population with the quality, nor at the original price.

But although this was in 2011, and it is assumed that the brands of oils that could have presented anomalies in their products no longer have them, because they would have taken the corresponding measures so that the fact is not repeated; the truth is that it continues raining on the wet in a despicable practice that is not only now, but has been many years, perhaps hundreds of years being carried out in a reprehensible and improper way.

Thus in the School encyclopaedic dictionary of 1889 it is stated that:

"The oils are the object of many adulterations, which consist of mixing with them inferior oils in quality and price, fats or animal oils, or also mineral oils, resin oils, fatty acids and even resins".

"The adulterations that are most interesting to know are those that are practiced with olive oil. This oil is adulterated or falsified by adding others of lower price, such as sesame, rapeseed, poppies, cotton and peanuts.

The examination carried out by the **OCU** in 2011 was carried out by taking samples of 40 virgin category olive oils in its two modalities: 34 extra virgin and 6 virgin, which were subjected to the corresponding physicochemical analyses, as well as the organoleptic tests established by qualified professionals.

The results obtained showed that in twelve of them (30 %) the consumer was deceived by offering him an oil of lower quality than the one labelled on the label. The report noted, however, that the health of the consumer was not put at risk by the type of anomaly detected, but by purchasing a type of product that was not the one he paid for or wanted to buy, with the consequent double effect: ethical and economic, something like buying cows and taking goats, or cats for hares.

The report in question detailed the type of oil, packaging, brand, supermarket chain of sale and the qualitative assessment of the body, to determine exactly which type was being sold. As for the tasting tests, emphasis was placed on organoleptic defects, fruitiness and fruit quality, which are what truly differentiate an extra virgin olive oil from a virgin to a dry one.

The laboratories chosen for the analyses are recognised by **ENAC** (Entidad Nacional de Acreditación, Spain) and the **COI** (International Olive Council) and have deserved prestige. The analysis was published in October 2012.

Although the report details each of the brands, this is not the subject of this paper, as it is such a widespread evil that it would take us a long time and space to detail it, but it must be noted that to be a study conducted by a neutral institution, on brands recognized as the world's leading producer of olive oil, it is very worrying that 30 % of the products valued do not meet the requirements laid down in the marketing standards, and that at least two were not suitable for sale and consumption, so they should have undergone a refining process and been offered as such.

The **OCU** report gave much to talk about, and it seems that the bodies involved took note of what happened, although some showed their disagreement by resorting to arguments, which rather than clear up doubts, did not favour their purposes at all, as they made them more evident; as well as a control system that, in the author's opinion, does not respond to the demands of the present time regarding an activity as delicate and important as food, which directly affects the health and welfare of citizens.

It is true that with time wounds are healed and memories are forgotten, but this case, in view of what has been explained above, is worrying, above all because of the disagreements of opinions of the conflicting sides, although the **OCU** has always been of the opinion that what was proved on one occasion does not have to be the same years later, especially if those involved decide to correct the errors, so the sentence should not be eternal, although no new controls have been carried out by this body.

Much more recent, however, and not related to the **OCU**, is that a new case of fraud was detected in Andalusia in which 120 tonnes of ready to mix oils were involved and converted into adulterated olive oil. The oils in question were sunflower oil, palm oil and avocado oil, which were to be mixed with appropriate flavourings and colourings to pass them off as virgin olive oil.

According to the authorities, the process was aborted in time, without adulterated products reaching the sales and distribution chain. But more than the fact of attempted adulteration, it is necessary to think that some mechanisms of analysis and control of the production and distribution chain are flawed, so that there is some possibility that once committed a fraud of this type the adulterated product can reach the market and then the consumer, because otherwise no one would engage in this type of fraud, under penalty of keeping the product undistributed.

The mixtures of oils of different nature present a considerable additional risk for the population, because besides being known by studies what properties would be altered, and not for the good of the consumer, two of these types of oils contain allergens for some people such as avocado and palm. In addition, when the adulterated product undergoes the heating process, either by cooking, or worse still by frying, its components could undergo changes in their properties and chemical composition with prejudice to people.

As if this were not enough, in April of 2106, the Government of Spain alerted the Autonomous Communities about the need to take extreme precautions and controls, because apparently, fraud was suspected to distort the quality of olive oils by some producers and pass them off as virgin extra virgin.

The fact itself has to do with the original samples sealed by inspectors that are later changed for better quality oils, violating the seal, so that they are considered to be extra virgin when they are of lower quality although the test once altered indicates otherwise. These seals, which should not be manipulated by anyone, except by the administration for appropriate controls and supervisions, seem to have been, and this has motivated the letter of the competent State agencies.

In its letter, the General Manager for the Agro-Food Industry argued that contradictory analyses had been detected in certain samples related to organoleptic tests of virgin olive oils, the results of which did not correspond to what the sealed samples show. In other words that the oil that had been monitored did not correspond to what was indicated in the sealed tests.

From the letter in question: *"An analytical follow-up has been carried out on the three samples analysed in a series of suspicious files and it has been objectively established that the analytical indicators of one of the samples do not coincide with those of the other two". "... "In all the cases discovered so far, the samples in the possession of the administration have provided the same analytical results, while these have been different in the sample in the possession of the administrator [the producing company].*

This fact is apparently not new, and already in 2011 the **FACUA (NGO** Consumers in Action, Spain) had condemned in Andalusia this type of practice in low-price brands because according to the agency, stains *"... the image above all of the cheapest firms in the market, giving the idea that the cheap is* *fraudulent," and calls for greater transparency in these problems as "...*only way to end the impunity of those who defraud".

According to the established regulations, it is always obligatory to collect three samples: initial for the laboratory, contradictory for the interested party and decisive for the public administration.

In this sense, *"When the initial one goes wrong, the contradictory one is analyzed. If it also goes wrong, there is a sanction. But if it goes well, you have to analyze the diriment because the operator could have manipulated it. They lift the seal and change the oil".* Agro-food sources from Aragón explain the 'modus operandi' of the fraudsters.

Fraud in the olive tree industry is not only adulterating quality; it has also splashed onto subsidies, and in 2002 *"through the olive operation"* aimed at combating the collection of fraudulent aid from the European Union to olive oil production, fraudulent facts were detected in a process that had begun since 1999, and which had numerous complaints. The fraud was estimated at over 3 million of euros, although more than 15 years have passed and there is no evidence that this situation continues.

Supermarket chains have also been involved in these unfortunate events, and the Junta de Andalucía ratified fraud in one of them in 2010, according to complaints from agricultural organisations that detected packages of white brand oil with labels that said they were *extra virgin* and were not; but blends of refined olive and virgin oils in a certain proportion, ie: anomalies in the labelling and in the nature of the product offered to the consumer.

Sometimes, among those apparently involved in the fraud there are attempts at deception by one or both parties, such as the recent fact that consignments of virgin olive oil sent to Italy from Spain, which, according to Spanish controls, did not bring together the indicators required and which were supposedly purchased to be packaged under private labels even if they were not. That is, double adulteration, or by both parties.

Apparently, the oils were sold in Italy by different suppliers who made the product go through "*de-acidification and deodorization of olive oil*", so that it "*became extra virgin olive oil*" or "*glaring virgin olive oil*", denominations by which it is exported to this country.

According to the report of the Italian authorities at the port of destination, the oils arriving in tankers as extra virgin olive oil gave '*irregular*' values at the time of the checks in six of the samples analysed, while in five others the analytical tests indicated anomalous values and the use of technical treatments not suitable for improving their quality.

Leaving the Old Continent, in the United States the New York Times has published that there is evidence that olive oils packaged under Italian brands such as extra virgin, high value in the market, themselves are olive oils from Spain, Tunisia and Morocco, which send them in ships or tankers to this country to be repackaged under their traditional brands.

The foregoing responds to the fact that, as expressed at the beginning, this is not only a problem of the countries of the Mediterranean Basin, since the University of California in the United States, in tests carried out in 2016, revealed that 70% of the analysed samples of olive oils sold in shops and supermarkets had been mixed with cheaper and lower quality oils (sunflower and canola).

This time the blame lay with United States (**US**) producers, who cheated consumers with a highly valued item on the market and a synonym, or product necessary for good health.

In Australia, in previous years, the government intensified controls and this determined that in 2012, none of the olive oils produced in this country obtained the category of *extra virgin*.

In South America, in Brazil, cases of adulteration have been detected not only in the oil imported from Europe, but also in the oil produced in this country, including oils that to a greater extent are not olive oils and are sold as such, being in some cases the market leaders.

In the same country, according to its Ministry of Agriculture, cases of fraud were indicated in olive oils imported mainly from Argentina, so they identified alterations in about 40% of the tests carried out in 45 oil brands in 13 states of the country. Brazil is one of the main importers of olive oil in the world, after the United States, and recently is making efforts to expand its domestic production.

In Uruguay, a report by the Food Sensory Evaluation Section of the Faculty of Chemistry of a university responsible for these controls indicates that most of the extra virgin olive oils marketed in the country do not correspond to their quality with the label. According to the report of this organism, most of the oils come from Argentina, Spain and Italy.

The previous controversial study in question was based on 19 samples from Uruguay, 12 from Spain, 6 from Argentina and 4 from Italy, taken at random from different commercial establishments. The result of the analysis showed that none of the foreign oils responded to the *extra virgin* quality, on the other hand, 84.5 of the Uruguayans, which gave some reasons for questioning, bearing in mind that in recent times there has been a strong tendency in the sector by emerging producers to defend internal interests in a somewhat sectarian way, blaming quality problems on imported products.

In this sense, if these doubts are sustained, this problem is not only Uruguay's own, and if we have any reference, it is also manifested in other countries that bid for their olive oil industry to compete and prevail over the others, such is the case of the United States, and also Brazil, although they do not have sufficient arguments to support this assertion, taking into account the subjective nature of the organosensory tests, which are now turned against those who endorse and establish them.

In Mexico, the magazine *Selecciones* (July 28, 2017) echoed the fraud of the packaging of non-original oil by traditional brands, taking as an example Italy where it is categorical to state: *"Italy produces some of the best oil in the world, but the labels on the bottles do not always tell the truth about their content."*

In a pleasant, somewhat novel way, the magazine echoes the visit of a correspondent to the Italian region of Apulia and the hardships of traditional producers of olive groves that have hundreds of years old, taking as an example that of an old olive grower who fought against wind and tide to maintain his small farm and oil mill, in a production that began their ancestors in times as early as the late sixteenth century by methods as traditional as a stone mill where: *"The juice is separated into water and oil in a centrifuge and, without using additives, heat or any refining process, it is transformed into extra virgin olive oil of a luminous golden green tone".*

These small traditional producers consider that the Italian olive oil that is sold in supermarkets is a fraud, and this is said in one of the main olive oil producing regions of the country, which contributes to the ruin and disappearance of these, despite the effort and work of the locals, because the olive trees are their history and passion, perhaps one of the main reasons for their long wanderings between olive groves.

According to the article, *"Cheap olive oil produced in Tunisia, Morocco, Spain and Greece is often relabelled as if it were Italian. That oil does not come from Italian olive groves, but at least it is made from olives."* To defend at least the origin of these against other false and adulterated oils containing other seed mixed with chlorophyll.

According to the article in question, *"extra virgin"* oil is often the cheapest, mixed with olive marc oil (obtained from the remains of the pulp and skin of olives through the use of solvents), treated, deodorized and flavoured with beta-carotene. Because the fraud has reached such a limit that it is offered in supermarkets at a price lower than the production price, which is indicative of the degree and intensity of adulteration.

A significant part of Apulia's high quality extra virgin olive oil, produced by small producers with careful semi-artisanal techniques, is not consumed or sold directly by them, and is sent to large factories, where it is mixed with other oils to disguise its lower quality.

For Italian olive oil specialists it is impossible, under current production conditions, to satisfy the high demand for this product with the right quality in accordance with the benefits required for health, and that hopeful consumers expect to acquire a valuable jewel of the Mediterranean diet. Behind this is, of course the interests of the large producers and above all their great influence on government decisions.

Tom Mueller's book **Extra Virginity: the Sublime and Scandalous World of Olive Oil**, which sheds light on this problem, is interesting. From to this author: *"Italy is at the forefront of scientific and research technology to be able to distinguish good products from bad"*.

In May 2015, an analysis carried out by the chemical laboratory of the Office of Customs and Monopolies in Rome found defects in 9 of 20 popular brands of oils in supermarkets. Analysis, which after repeated, yielded the same results in traditional Italian oil brands, whose test did not agree with the labelling. There were also important supermarket chains affected by or concurrent with this problem.

As in the case of Spain with the **OCU**, the Italian producing companies, some of them even repeat offenders in this type of problem, denied having incurred in anomalies and questioned the reliability of the taste tests. Their arguments were rejected by the representatives of the **AGCM** (*Autorità garante della concorrenza e del mercato*), although they were entitled to pursue their appeals.

These problems are repetitive in Italy and the authorities recently withdrew some 2 000 MT of extra virgin olive oil, apparently of Greek and Spanish origin, sold under a prestigious Italian brand in Italy on the basis of false documentation, which was then sent to the factories to be packaged and marketed inside and outside the country.

According to the Italian Association of Farmers *"Coldiretti"*, it is worrying: *"the large number of frauds with imported olive oils that are frequently mixed with those originating in the country to obtain a label of Italian origin that harms Italian producers and consumers"*.

The problems of oil adulteration, which, as we have seen, reach as far as Oceania, also affect Asian countries, and in 2013 the Taiwanese authorities withdrew several batches of olive oils from Spain, but processed on this island, due to the suspicion that they were adulterated with *chlorophylline*, a food additive, E140ii, which is a semi-natural dye obtained by treating chlorophyll with copper and which is soluble in water, does not show toxicity at concentrations indicated as an additive, but does release this metal by heating.

The Taiwanesis authorities at the factory where the fraud was investigated requisitioned oils and documents related to the formulations used by the factory in processing the oils.

According to the analyses carried out, the original virgin oil from Spain was adulterated and later sold as such, although in reality it was a mixture with other cheaper oils, and *chlorophylline* was used to encourage colour and make it look like extra virgin olive oil. From the requisite documentation it was deduced that the adulterated oil in question could contain up to 20 % *camellia oil*.

Initially, high levels of acidity were also detected in other olive oils, although the tests are not conclusive. As in other countries, Taiwanese producers initially justified the anomalies as due to problems in the bottling process.

New analysis tools for olive oils.

Considering the previously valued aspects, and as most of the virgin olive oil marketed in the world is produced in Europe, as well as the fact that adulteration problems affect the prestige and the economy of a traditional sector with strong cultural roots; new techniques are currently being studied to analyse and detect as quickly and safely as possible the quality and originality of olive oils in order to avoid, hinder and unmask possible fraudulent acts.

Within these techniques is the use of **DNA** that allowed verify that olive oils from a consortium in Tuscany showed documents of originality of olive oils, which in the tests did not correspond to those obtained in that region, but came from lots imported from Greece and Apulia.

Work is also being done on the introduction of intelligent labels with codes on the origin and distribution network of the oils. Nanotechnology is also being used using magnetic **DNA** nanoparticles in experiments carried out by Swiss researchers.

In particular, in olive oil, the use of **NIR** spectroscopy (Near Infrared) seems to be an appropriate method that allows, by means of classical spectrophotometric methods, to accurately determine the components under analysis, according to a technique proposed by food specialist Dr. Christian Gertz, for the ultra-fast determination of polar compounds, taking advantage of their radiation absorption in the spectral regions between the visible and the infrared.

BIBLIOGRAPHY RELATED TO THE SUBJECT CONSULTED ON THE NETWORK.

-Aceite de oliva falso: Peligros del aceite de oliva adulterado y marcas …
www.saludcasera.com/grasas.../aceite-de-oliva-falso-adulterado-marcas-evitar (2016).

-Adulteraciones en el aceite de oliva | Catalunya Vanguardista.
www.catalunyavanguardista.com/catvan/adulteraciones-en-el-aceite-de-oliva/ (7 marz. 2011)

-ABS ECONOMÍA. http://www.abc.es/economia/abci-italia-retira-mas-2000-toneladas-aceite-espanol-y-griego-vendidas-como-producto-

-Calidad del aceite de oliva. Fraude en el aceite de oliva | MeetSpain
www.meetspain.es/.../polemica-infografia-de-new-york-times-sobre-el-aceite-de-oliva.

-Confiscan 2 mil toneladas de aceite de oliva italiano adulterado
www.notitarde.com/Internacional/...aceite-de-oliva-italiano-adulterado--/.../873558/ (3 feb. 2016).

-Detección de adulteraciones en aceites de oliva vírgenes ... - Expoliva
www.expoliva.com/expoliva2005/simposium/comunicaciones/TEC-51.pdf

-Detección de adulteraciones o contaminaciones del aceite de oliva ...
grasasyaceites.revistas.csic.es/index.php/grasasyaceites/article/download/496/498 (2008).

-Desarticulada una red que vendía aceite de oliva adulterado | Edición ...
elpais.com › Sociedad (12 ab. 2016).

-El ataque a la 'marca España' patina con el aceite de oliva | Andalucía ...
www.elmundo.es/elmundo/2012/10/05/andalucia/1349454849.html (5 oct. 2012),

-El gran fraude del aceite de oliva - Revista Selecciones, México
https://mx.selecciones.com/la-gran-estafa-del-aceite-de-oliva/.
(23 de julio de 2017). Sorrel Sorrel Downer. Con investigación y textos adicionales de Lorraine Shah.

-El aceite adulterado intervenido en La Bañeza tenía efectos nocivos pa..
www.lanuevacronica.com/el-aceite-adulterado-intervenido-en-la-baneza-tenia-efectos... (19 oct 2015).

-Escolar.com (Diccionario Enciclopédico 1889). 2014 - Diccionario Enciclopédico Hispano-Americano Siglo XIX. ACEITE. Adulteraciones del aceite de oliva.
ACEITE - Adulteraciones del aceite de oliva - Escolar.com
www.escolar.com/EnciclopediaXIX/aceite-oliva-adulteraciones.html

-El Gobierno detecta un fraude para falsear la calidad del aceite de ...
www.eldiario.es/sociedad/Agricultura-detecta-fraude-controles_0_507449640.html (22 abril 2016)

-EE.UU perseguirá la adulteración del aceite de oliva con aceite de ...
www.olimerca.com/...adulteracion...aceite-de-liva...aceite.../d85e13456205d1b93a0...(1 de jun. 2016).

-EEUU: la FDA podría revisar las normas de autenticidad para el aceite ...
www.portalolivicola.com › Internacionales (27 jul. 2017).

-El fraude alimentario – Clarín.
https://www.clarin.com/buena-vida/.../fraude-

alimentario_0_H1ulGOojwQl.html
(29 enero 2013).

-Fraude del aceite de oliva en Estados Unidos | Gastronomía &
Cía.
https://gastronomiaycia.republica.com/.../fraude-del-aceite-de-
oliva-en-estados-unidos... (2 de nov. 2010).

-Fraude en el aceite de oliva virgen extra de España |
Gastronomía & Cía
https://gastronomiaycia.republica.com/.../fraude-en-el-aceite-de-
oliva-virgen-extra-de...(24 nov. 2010).

-Fraude del aceite de oliva en Uruguay | Gastronomía & Cía
https://gastronomiaycia.republica.com/2010/.../fraude-del-
aceite-de-oliva-en-uruguay/ (3 nov. 2010).

-Gastronomiaycia.republica.com/2010/11/03/fraude-del-aceite-
de-oliva-en-uruguay/

-https://okdiario.com/.../agricultura-analiza-si-dia-adultera-
aceite-oliva-mezclandolo-gi. (15 de jun. 2016)

-Investigan a una empresa de Córdoba por la supuesta venta de
aceite ...
sevilla.abc.es/.../sevi-investigan-empresa-supuesta-venta-aceite-
adulterado-italia-2017. (11 feb. 2017).

-Italia 'destapa' un fraude del aceite de oliva español vendido
como ...
www.eleconomista.es/.../Italia-destapa-el-fraude-del-aceite-
espanol-vendido-como-ital.. (3 feb 2006).

-INMETRO analiza marcas de aceite de oliva extra virgen en
Brasil.
www.revistaquimica.cl/?p=3861 (9 enero 2016).

-La estafa del aceite de oliva virgen italiano - Cocinillas
cocinillas.elespanol.com/2014/01/la-estafa-del-aceite-de-oliva-

virgen-italiano/ (31 enero 2014).

-Los carabineros desarticulan una red que comercializaba aceite de ...
www.publico.es/actualidad/carabineros-desarticulan-red-comercializaba-aceite.html (21 feb 2011).

-Mueller, T. (2011): *Extra Virginity. The Sublime and Scandalous World of Olive Oil* W.W. Norton and Company. ISBN: 0393070212

-Nuevos marcadores de adulteraciones en aceites de oliva.
noticias.universia.es › Noticias › Portada (10 marz. 2011).

-Nuevos pasos contra el fraude del aceite de oliva en Brasil - Icex
www.icex.es/icex/es/Navegacion-zona-contacto/revista-el.../NEW2017712216.html (7 may. 2017).

-OCU-Compra Maestra nº 375 Noviembre 2012. www.ocu.org

-Trazabilidad y Autentificación de Aceites.
https://www.trafoon.org/sites/trafoon.org/files/jaen_garcia_2015 05.pdf (6 may. 2015)

-Técnicas basadas en ADN para detectar aceite de oliva adulterado ...
www.agenciasinc.es/.../Tecnicas-basadas-en-ADN-para-detectar-aceite-de-oliva-adulte..(17 may 2017)

-Taiwan retira un aceite de oliva español adulterado por una empresa ...
www.elmundo.es/elmundo/2013/10/18/economia/1382059146.html (18 oct. 2013)

-Tecnología alimenticia para luchar contra fraudes como el del aceite ...
https://www.technologyreview.es/.../tecnologia-alimenticia-para-luchar-contra-fraudes.. (26 may 2015)

-The New York Times desata la polémica en torno al aceite de oliva ... - Icex
www.icex.es/icex/es/navegacion-principal/todos-nuestros...de.../4729675.html?...(2017)

-Taiwan denuncia que un Aceite de Orujo de Oliva Italiano contiene ...
andexconsultores.es/taiwan-denuncia-que-un-aceite-de-orujo-de-oliva-italiano-contie... (16 enero 2014)

-Universidad de California revela las marcas que venden aceite de ...
tlvz.com/marcas-que-venden-aceite-de-oliva-falso/ 2016

CHAPTER ANNEX

Some standards contained in the OC/T.15/NC bulletin no. 3/Rev. 7 May 2013 referring to olive oils and marc.

COMMERCIAL STANDARD APPLICABLE TO OLIVE OILS AND OLIVE-MARC OILS

1. SCOPE OF APPLICATION

This rule shall apply to olive oils and olive-marc oils which are the subject of international trade or of transactions in the form of concessions or food aid.

2. NAMES AND DEFINITIONS

Olive oil is oil obtained solely from the fruit of the olive tree (*Olea europaea L.*), to the exclusion of oils obtained by solvents or by re-esterification processes and any mixture with oils of other kinds. It shall be marketed in accordance with the following designations and definitions:

2.1.1. Virgin olive oils are oils obtained from the fruit of the olive tree solely by mechanical or other physical means under conditions, particularly thermal conditions, which do not lead to alteration of the oil, which has not undergone any treatment other than washing, decantation, centrifugation and filtration.

2.1.1.1. Virgin olive oils fit for consumption in the form in which they are held are included:

(i) extra virgin olive oil: virgin olive oil having a free acidity expressed as oleic acid of not more than 0,8 grams per 100 grams and the other characteristics of which correspond to those laid down for this category in this standard.

(ii) <u>virgin olive oil</u>: virgin olive oil, the free acidity of which, expressed as oleic acid, is not more than 2 grams per 100 grams and the other characteristics of which correspond to those laid down for this category in this Standard.

(iii) ordinary virgin olive oil: virgin olive oil having a free acidity expressed as oleic acid of not more than 3,3 grams per 100 grams and the other characteristics of which correspond to those laid down for this category in this Standard.

2.1.1.2.. <u>Virgin olive oil</u> <u>not fit for consumption in the form in which it is obtained, called glaring virgin olive oil</u>: virgin olive oil, the free acidity of which, expressed as oleic acid, is greater than 3.3 grams per 100 grams and/or the organoleptic characteristics and other characteristics of which correspond to those laid down for this category in this Standard. It is intended for refining industries or technical uses.

2.1.2 . <u>Refined olive oil</u> is olive oil obtained from virgin olive oils by means of refining techniques which do not lead to any modification of the initial glyceridic structure. Their free acidity expressed as oleic acid is a maximum of 0.3 grams per 100 grams and their other characteristics correspond to those fixed for this category in this Standard.

2.1.3<u>. Olive oil</u> is the oil made up of a mixture of refined olive oil and virgin olive oils fit for consumption in the form in which they are obtained. Their free acidity expressed as oleic acid is a maximum of 1 gram per 100 grams and their other characteristics correspond to those laid down for this category in this Standard.

2.2. **Olive-marc oil** is the oil obtained by treating olive marc with solvents or other physical processes, to the exclusion of oils obtained by re-esterification processes and any mixture with oils of other kinds. It shall be marketed in accordance with the following designations and definitions:

Crude olive-marc oil is olive-marc oil the characteristics of which correspond to those laid down for this category in this standard. It is intended for refining for human consumption or technical use.

2.2.2. Refined olive-marc oil is oil obtained from crude olive-marc oil by refining techniques which do not result in any modification of the initial glyceridic structure. Its free acidity expressed as oleic acid is a maximum of 0,3 grams per 100 grams and its other characteristics correspond to those laid down for this category in this standard.

2.2.3. Olive-marc oil is the oil made up of a mixture of refined olive-marc oil and virgin olive oil fit for consumption in the form in which it is obtained. Its free acidity expressed as oleic acid is a maximum of 1 gram per 100 grams and its other characteristics correspond to those laid down for this category in this Standard. This mixture may under no circumstances be called "olive oil".

3. PURITY CRITERIA

The identification characteristics constituting the purity criteria are applicable to olive oils and olive-marc oils.

The limits established for each criterion include the precision margins of the recommended method.

Fatty acid composition by gas chromatography (% m/m of methyl esters)

- Myristic acid ≤ 0,03
- Palmitic acid 7,50 - 20,00
- Palmitoleic acid 3,30 - 3,50
- Heptadecanoic acid ≤ 0,30
- Heptadecenoic acid ≤ 0,30
- Stearic acid 0,50 - 5,00
- Oleic acid 55,00 - 83,00
- Linoleic acid 3,50 - 21,00
- Linolenic acid ≤ 1,00
- Arachidic acid ≤ 0,60
- Gadoleic acid (eiosenoic) ≤ 0,40
- Behenic acid ≤ 0,20
- Lignoceric acid ≤ 0,20

Trans-fatty acid content (% of trans-fatty acids)

	C18:1 T	C18:2 T + C18:3 T
	%	%
-Edible virgin olive oils	≤ 0,05	≤ 0,05
-Glaring virgin olive oil	≤ 0,10	≤ 0,10
-Refined olive oil	≤ 0,20	≤ 0,30
-Olive oil	≤ 0,20	≤ 0,30
-Crude olive marc oil	≤ 0,20	≤ 0,10
-Refined olive marc oil	≤ 0,40	≤ 0,35
-Olive marc oil	≤ 0,40	≤ 0,35

3.3 Composition in sterols and triterpenic dialcohols

3.3.1 Composition in desmethylsterols (% of total sterols)

- Cholesterol $\leq 0,5$
- Brasicasterol $\leq 0,1$
- Campesterol $\leq 4,0$
- Stigmasterol $<$ campesterol for edible oils
- Delta-7-stigmastenol $\leq 0,5$
- Betasitosterol aparente:
  - betasitosterol +
  - delta-5-avenasterol +
  - delta-5-23-estigmastadienol +
  - clerosterol + sitostanol +
  - delta-5-24-stigmastadienol ≥ 93

3.3.2. Content in total sterols (mg/kg)

-

- - Virgin olive oils ≥ 1000

- - Refined olive oil ≥ 1000

- - Olive oil ≥ 1000

- - Crude olive marc oil ≥ 2500

- - Refined olive marc oil ≥ 1800

- - Olive marc oil ≥ 1600

-

3-3.3 Erythrodiol and uvaol content (% of total sterols)

- Edible virgin olive oils $\leq 4,5$

- Glaring virgin olive oil $\leq 4,5$

- Refined olive oil $\leq 4,5$

- Olive oil $\leq 4,5$

- Crude olive-marc oil $> 4,5$

- Refined olive-marc oil $> 4,5$

- Olive marc oil $> 4,5$

3.4. Wax content

C42 + C44 + C46 (mg/kg)

- Extra virgin and edible virgin olive oils ≤ 150
- C40 + C42 + C44 + C46 (mg/kg)
- Current olive oil ≤ 250
- Glaring virgin olive oil ≤ 300
- Refined olive oil ≤ 350
- Olive oil ≤ 350
- Crude olive-marc oil > 350
- Refined olive-marc oil > 350
- Olive marc oil > 350

3.5. Maximum difference between the actual content and the theoretical content in triglycerides with ECN 42

- Edible virgin olive oils $\leq \left| 0,2 \right|$
- Glaring virgin olive oil $\leq \left| 0,3 \right|$
- Refined olive oil $\leq \left| 0,3 \right|$
- Olive oil $\leq \left| 0,3 \right|$
- Crude olive marc oil $\leq \left| 0,6 \right|$
- Refined olive marc oil $\leq \left| 0,5 \right|$
- Olive marc oil $\leq \left| 0,5 \right|$

3.6. Content in stigmastadienes (mg/kg)

- Extra virgin and virgin olive oils $\leq 0,05$

- Edible virgin olive oils ≤ 0,10
- Glaring virgin olive oil ≤ 0,50

3.7. Content of 2 glyceryl monopalmitate

-Edible virgin olive oils and olive oil

$C16:0 \leq 14.0\%$; $2P \leq 0.9\%$
$C16:0 > 14.0\%$; $2P \leq 1.0\%$

- Non-edible virgin olive oils and refined olive oils
$C16:0 \leq 14.0\%$; $2P \leq 0.9\%$
$C16:0 > 14.0\%$; $2P \leq 1.1\%$

-Olive marc oils ≤ 1.2 %.
-Crude and refined olive-marc oils ≤ 1.4 %

Unsaponifiable matter (g/kg)

-Olive oils ≤ 15

-Olive marc oils ≤ 30

5. FOOD ADDITIVES

5.1. Virgin olive oils and crude olive-marc oil: no additives are permitted.

5.2. Refined olive oil, olive oil, refined olive-marc oil and olive-marc oil: alpha-tocopherol authorised to restore the natural tocopherol lost during refining.

Maximum dose: 200 mg/kg total alpha-tocopherol in the final product.

6. CONTAMINANTS

6.1 Heavy metals

The products covered by the provisions of this Standard shall comply with the maximum limits for heavy metals established by the Codex Alimentarius Commission, but in the meantime the following limits shall apply to them:

Maximum Allowable Concentration

Lead (Pb) 0,1 mg/kg
Arsenic (As) 0,1 mg/kg

6.2 Pesticide residues.

The products covered by this Standard shall comply with the maximum residue limits established by the Codex Alimentarius Commission for these products.

6.3 Halogenated solvents

- Maximum content of each halogenated solvent 0,1 mg/kg

- Maximum content of total halogenated solvents 0,2 mg/kg

Other Works by the Author

CONFUCIO PARA CONFUSOS

480 AFORISMOS DE CONFUCIO

CALIXTO LÓPEZ

THE ETHICAL AND MORAL CODE OF CONFUCIUS

Calixto López
Rosalía Rouco

EL PELIGROSO ARTE DE FREÍR

CALIXTO LÓPEZ
ROSALÍA ROUCO

QUÍMICA DE LOS ACEITES VEGETALES

CALIXTO LÓPEZ HERNÁNDEZ

ACEITE DE COCO

CALIXTO LÓPEZ HERNÁNDEZ

COCONUT OIL DIET AND OBESITY

CALIXTO LÓPEZ HERNÁNDEZ

BIBLIOGRAPHY

Abbott, A. (2017): *Italy rebuked for failure to prevent olive-tree tragedy*1 Nature (546); pp. 193-194.

Alba J. y L. Martínez. (2001). *Elaboración de Aceites de Oliva*. En: Mataix J, editor.

Ancin, M. y M. Martinez. (1991). *Estudio de la degradación de los aceites de oliva sometidos a fritura.* Ácidos y Grasas. 1991 (1), (42): 22-31.

Almeida, R. and L. Nunney. (2015): *How do plant diseases caused by Xylella fastidiosa emerge?*; Plant Disease (99); 1457-1467.

Anderson, J., F. Grande and A. Keys (1970). *Coronary heart disease in Seven countries"*. Circulation, 1970, 41; 1-211

Angerosa F, et al. (2004). (2004*). Volatile compounds in virgin olive oil: occurrence and their relationship with the quality.* J Chromatogr A 2004; 1054: 17-31.

Aparicio, R, y J. Harwood. (2003). *Manual del Aceite de Oliva*. AMV Ediciones y Mundi-Prensa. Madrid. 2003.

Arbonés-Mainar J. (2008). *Olive oil phenolic compounds as potential therapeutical agents.* La Veletta: Nova; 2008.

Astiasarán, Y. y J. Martínez, (2003). *Alimentos. Composición y propiedades*. McGraw-Hill Interamericana. Madrid.

AOCS. (1997). *Official Methods and Recommended Practices of the American Oil* Chemists Society, 5th ed. D. Firestone (ed), AOCS Press, Champaign.

Ávila, J. (2000). *Enciclopedia Del Aceite De Oliva*. 1° Ed. Editorial Planeta. Barcelona, España.

Badui, S. (2006) *Química de los alimentos*. 4ta. Edic. PEARSON. Adison Wesley. México.

Bailey, A. (1998), *Aceites y grasas industriales*, Editorial Reverte, España.

Bailey, A. (1961). *Química de los Alimentos*. 3ra. ed. Editorial Addison Wesley Longman. México.

Bastida S., and F. Sánchez-Muniz. (2001). *Thermal oxidation of olive oil, sunflower oil and a mix of both oils during forty discontinuos domestic frying of different foods*. Food Sci Tech Int. 2001; 7: 15-21.

Beauchamp, G. et al. (2005). *Ibuprofen-like activity in extra-virgin olive oil*. Nature, 2005, 437, 45-6).

Blekas, G., M. Tsimidou and D. Boskou. (1995). *Contribution of α-tocopherol to olive oil stability. Food chemistry* 52 (3): 289-294.

Barranco, D. (1995). *La Elección Varietal en España*. Ed. Consejo Oleícola Internacional. Olivae 59. 54-58 1995.

Berra, B. (1998). *Biochemical and nutricional aspect of the minor component of olive oil*. Olivae. 73, 29-30.

Boskou, D. (2006). *Sources of natural phenolic antioxidants*. Trends in Food Science and Technology. 17 (9): 505-512

Boskou D. (1998). *Olive oil, Chemistry and Technology*. AOLS Press: Champaing ed; 1998.

Brenes M, et al. (1999). Phenolic *compounds in Spanish olive oils*. J Agric Food Chem 1999; 47:3535-40.

Brenes M, et al. (2002). *Influence of thermal treatments simulating cooking processes on the polyphenol content in virgin olive oil*. J Agric Food Chem. 2002; 50: 5962-5967.

Coultate, T. (1998). *Manual de Química y Bioquímica de los alimentos*. Ed Acribia. España.

Cicerale, S. et al. (2009). *Chemistry and health of olive oil phenolics*. Critical Reviews in Food Science and Nutrition 49: 218–236.

Castillo, E. Torres, S. y B. Álvarez (2007). *El aceite de oliva y la salud. Proceso industrial y puntos críticos de control en almazaras.* Higiene y Sanidad Ambiental, 7: 256-264 (2007).

Civantos L., Contreras R. y R.Grana. (1999). *Obtención del Aceite de Oliva Virgen.* Madrid: Editorial Agrícola Española, 2ª Edición; 1999.

Covas M. (2007*). Olive oil and the cardiovascular system.* Pharmacol Res 2007; 55: 175-186.

Covas, M., et al. (2006). *The effect of polyphenols in olive oil on heart disease risk factors: a randomized trial.* Ann Intern Med 2006;145: 333-141.

Covas, M et al. (2006). *Minor components of olive oil. Evidence to date of health benefits in humans.* Nutr Rev; 64(Suppl 1):20-30.

Comisión Europea. «REGLAMENTO (CE) No 1513/2001 DEL CONSEJO de 23 de julio de 2001 que modifica el Reglamento no 136/66/CEE y el Reglamento (CE) no 1638/98, en lo que respecta a la prolongación del régimen de ayuda y la estrategia de la calidad para el aceite de oliva».

Consejo Oleícola Internacional. Norma Comercial Aplicable al Aceite de Oliva y al Aceite de Orujo. 17. (1998).

Clevidence B, et al. (1997). *Plasma lipoprotein (a) levels in men and women consuming diets enriched in saturated, cis-, or transmonounsaturated fatty acids*. Arterioscler Thromb Vasc Biol 1997; 17: 1657-61.

Carmena R, et al. (1996). *Effect of olive and sunflower oils on low density lipoprotein level, composition, size, oxidation and interaction with arterial proteoglycans*. Atherosclerosis 1996;125:243-255.

Del Castillo, E., V. Torres y B.Álvarez. (2007). El aceite de oliva y la salud. Proceso industrial y puntos críticos de control en almazaras Hig. Sanid. Ambient. 7: 256-264.

Denance, N. et al. (2015). *Several subspecies and sequence types are associated with the emergence of Xylella fastidiosa in natural settings in France*; Plant Pathology (66); 1054-1064.

Departamento de Salud y Servicios Sociales de los Estados Unidos (2010). Dietary Guidelines for Americans.

Dudrow F. (1983). *Deodorization of edible oil*. J. of Am. Oil. Chem. Soc. 60, 272-274.

Espínola, F. (1996). *Cambios tecnológicos en la extracción del aceite oliva virgen*. Alimentación, equipos y tecnología 1996.

Ferrari R. et al. (1996). *Minor constituents of vegetable oils during industrial processing*. J. Am. Oil Chem. Soc. 73, 587-591.

Fit, M. et al. (2007*). Bioavailability and antioxidant effects of olive oil phenolic compounds in humans: a review*. Ann IstSuper Sanita 2007; 43: 375-381.

Fito, M., et. al. (2008). *Anti-inflammatory effect of virgin olive oil in stable coronary disease patients: a randomized, crossover, controlled trial*. Eur J Clin Nutr 2008; 62:570

Foster A, and A. Harper A. (1983). *Physical refining*. J of Am Oil Chem. Soc. 60, 265-271.

Foster, R.; C. Williamson, and J. Lunn, (2009). *Culinary oils and their health effects* Nutrition Bulletin 34 (1): 4-47.

Frankel, E. (2011). *Nutritional and biological properties of extra virgin olive oil*. J. Agric. Food. Chem. 2011. 59 (3): 785-92.

Galli C and F. Visioli (1999). *Antioxidant and other activities of phenolics in olives/olive oill, typical components of the Mediterranean diet*. Lipids 1999; 34 23-26.

Garrido, J. et al.(1990). *Pigmentos clorofílicos y carotenoides responsables del color del aceite de oliva virgen*. Grasas y Aceites. 41. 2. 404-409. 1990.

Gandul, B.and M. Mínguez (1996). *Chlorophyll and carotenoid composition in virgin olive oils from various Spanish olive varieties*. J Am Oil Chem Soc, 72: 31-39.

Harwood J. and R Aparicio. (2000). *Handbook of olive oil, Analysis and Properties*: Kluwer Academic Publishers J.L. Harwood and R. Aparicio; 2000.
Hendrix, B. (1990). *Edible Fats and Oils Processing: Basic Principles and Modern practises*. Illinois: Ed. D.R. Erickson., Am.Oil Chem. Soc. Chamaing; 1990.

Horton J, et al. (1993). *Dietary fatty acids regulate hepatic low density lipoprotein (LDL) transport by altering LDL receptor protein and mRNA levels*. J Clin Invest 1993; 92: 743-49.

Hu, F., et al. (1997). Dietary fat intake and risk of coronary heart disease in women. N Engl J Med 1997; 337: 1491-99.

https://pixabay.com/es/

James, C. (1996). *Analytical Chemistry of Foods*. Blackie Academic and Professional. London.

Jiménez, A., et al. (1995). *Elaboración del aceite de oliva virgen mediante sistema continúo de dos fases: Influencia de las diferentes variables del proceso en algunos parámetros relacionados con la calidad del aceite*. Grasas y Aceites *(*46): 299-303.

Kamal-Eldin A, and L. Appelqvist (1996). *The chemistry and antioxidant properties of tocopherols and tocotrienols*. Lipids. 31, 671-701.

Keys A, J. Anderson and F. Grande (1957). *Prediction of serum cholesterol responses of man to changes in fats in the diet*. Lancet 1957; 273: 959-66.

Keys A. (1980). "*Seven Countries: A Multivariate Análisis of Death and Coronary Heart Disease*." Cambridge, MA: Harvard University Press.

Keys A., A. Mennoti, M. Karvonen C. Aravanis, H. Blackburn , et al. (1986) *The diet and 15-year death rate in the seven countries study*. Am J Epidemiol 1986; 124: 903-915.

Khalil M, W. Wagner and I. Goldberg. (2004). *Molecular interactions leading to lipoprotein retention and the initiation of atherosclerosis*. Arterioscler Thromb Vasc Biol; 24: 2211-18.

Kris-Etherton P, and S. Yu (1997). *Individual fatty acids on plasma lipids and lipoproteins: human studies*. Am J Clin Nutr 1997; 65: 1628S-44S.

Kritchevsky D. (1998). *History of recommendations to the public about dietary fat*. J. Nutr 1998; 128: 449-52.

Kushi, L, et al. (1985) *Diet and 20-year mortality from coronary heart disease*. The Ireland-Boston Diet Diet-Heart Study. N Engl J Med 312: 811-8.

Lanzón, A, T; Cert and J. Gracián, (1994). *The hydrocarbon fraction of virgin olive oil and changes resulting from refining* Journal of the American Oil Chemists' Society 1994;71:285-291.

Lichtenstein A, et al.(2006). *Summary of American Herat Association diet and lifestyle recommendations revision.* Arterioscler Thromb Vasc Biol 2006; 26: 2186-91.

Loconsole, G. et al. (2016). *Intercepted isolates of Xylella fastidiosa in Europe reveal novel genetic diversity*1 ; J. Plant Pathol. (146); 85-94.

Lou-Bonafonte, J. el at. *Efficacy of bioactive compounds from extra virgin olive oil to modulate aterosclerosis development.* Mol. Nutr. Food Res. 2012, 56, 1043–1057.

López, C. (2018). *Aceites Vegetales.* Amazon Kindle Publishing ISBN.9781980870401. Spain.

López, C. (2017). *Caos e Incertidumbre en el Mundo de los Aceites Vegetales.* Amazon Kindle KDP Publishing, 9751549915190. Spain.

López, C. (2018). *El Peligroso Arte de Freir.* Amazon Kindle KDP Publishing. ISBN 9781973324423. Spain.

Montedoro, G, et al. (1992*). Simple and hydrolysable phenolic compounds in virgin olive oil: Their extraction, separation and quantitative and semiquantitative evaluation by HPLC.* J Agric Food Chem 1992;40:1571-1576.

Montedoro G. et al. (1998). *Antioxidants in virgin olive oil.* Olea 2007; 26: 5-13.

Mateos, R., et al. (2001). *Determination of phenols, flavones and lignans in virgin olive oil by soil-phase extraction and high-performance liquid chromatography with diodearray ultraviolet detection*. J Agri Food Chem 49: 2185-2219

Mataix J., J. Ochoa and J. Quiles. (2004). *Olive oil, dietary fat and ageing, a mitochondrial approach*" Grasas y Aceites Vol. 55. Fasc. 1 (2004), 84-91

Mattson F. and S. Grundy (1995). *Comparison of effects of dietary saturated, monounsaturated and polyunsaturated fatty acids on plasma lipids and lipoprotein in man*. J. Lipid Res., 26, 194-202.

Mensink, R. and M. Katan. (1992). *Effect of dietary fatty acids on serum lipids and lipoproteins. A metaanalysis of 27 trials*. Arterioscler Throm 12: 911-919, 1992.

Mensink R, et al. (2013). *Effects of dietary fatty acids and carbohydrates on the ratio of serum total to HDL cholesterol and on serum lipids and apolipoproteins: a meta-analysis of 60 controlled trials*. Am J Clin Nutr. (77) (5) pp.1146-1155.

Mozaffarian D, R. Clarke (2009). *Quantitative effects on cardiovascular risk factors and coronary heart disease risk of replacing partially hydrogenated vegetable oils with other fats and oils*. Eur J Clin Nutr 2009; 63: S22-S33.

Moreiras O. et al.(2007). *Tablas de composición de alimentos*. 11ª edición. Pirámide. Madrid.

Olmo, D. et al. (2017). *First detection of Xylella fastidiosa infecting cherry (Prunus avium) and Polygala myrtifolia plants, in Mallorca Island, Spain*1 Plant Dis. (101); 1820.

Owen, R. et al, (2000). *Phenolic compounds and squalene in olive oils: the concentration and antioxidant potential of total phenols, simple phenols, secoiridoids,*
lignansand squalene. Food Chem Toxicol 2000;38:647-59.

Organización Mundial de la Salud (2015). Avoiding Heart Attacks and Strokes. **Reglamento (CE) Nº 1989/2003 DE LA**

COMISIÓN de 6 de Noviembre de 2003, que modifica el Reglamento (CE) n° 2568/91, relativo a las características de los aceites de oliva y de los aceites de orujo de oliva y sobre sus métodos de análisis.

Parkinson, L. and R. Keast (2014) *Oleocanthal, a phenolic derived from virgin olive oil: a review of the beneficial effects on inflammatory disease.* Int. Journal of Molecular Sciences, 2014, 15, 12323-12334.

Palacio-Bielsa, A. (2017). *'XylellaFastidiosa', un problema global: enfermedades que causa, diagnóstico y control.* Centro de Investigación y Tecnología Agroalimentaria. Aragón, España.

Pellegrini N. et al. (2001). *Direct analysis of total antioxidant activity of olive oil and studies on the influence of heating.* J Agric Food Chem. 2001; 49: 2532-2538.

Pérez-Jiménez, F et al. (2007). *The influence of olive oil on human health: not a question of fat alone"* Mol. Nutr. Food Res. 2007, 51, 1199 – 1208

Quiles, J., et al. *Olive Oil & Health.* (2006). CABI, Wallingford, UK. 2006.

Rigacci, S; Stefani, M (2016). *Nutraceutical Properties of Olive Polyphenols. An Itinerary form Cultured Cells through animal Models to Humans.* International Journal of Molecular Sciences. 17 (6): 843.

Smith, Amos, et al. (2005). *Síntesis y Asignación de Configuración Absoluta de (-)-Oleocantal: Potente Antioxidante No esteroide Antiinflamatorio Derivado de Aceites Extra Virgen de Oliva." . Organic Letters* (2005), 7(22), 5075-5078.

Sánchez Muñiz, F. and S. Bastida. (2006): *Effect of frying and thermal oxidation on olive oil and food quality, en Olive Oil and Health.* Quiles, J. M. Ramírez-Tortosa, P. Yaqoob (eds.) CAB

International, Oxfordshire, UK, 74-108.

Tarrago-Trani, M et al. (2006). *New and existing oils and fats used in products with reduced trans-fatty acid content..* Journal of the American Dietetic Association. pp. 867-880.

Verleyen T, et al. (2002). *Analysis of free and esterified sterols in vegetable oils.* J. Am. Oil Chem. Soc. 79, 117-122.

Vicent, A. and J. Blasco. (2017). *When prevention fails. Towards more efficient strategies for plant disease eradication* New Phytol. (214); 905-908;

Williams, C. et al. (1999). *Cholesterol reduction using manufactured foods high in monounsaturated fatty acids, a randomized cross-over study.* Br. J. Nutr., 81, 439-446.

Warner K, and N. Michael-Eskin (1995). *Methods to asses quality and stability of oils and fat-containing foods.* AOCS Press. Illinois, USA. Cap. 2,9.

Zschau W. (2000). *Introduction to Fats and Oils Technology,* 2nd edn. Champaign, IL: AOCS Press.

INDEX